Lectures on Mathematics in the Life Sciences
Volume 9

Some Mathematical Questions in Biology, VIII

The American Mathematical Society
Providence, Rhode Island
1977

Proceedings of the Tenth Symposium on
Mathematical Biology held in Boston, February, 1976.

edited by
Simon A. Levin

Library of Congress Catalog Card Number 77-25086
International Standard Book Number 0-8218-1159-2
AMS 1970 Subject Classification: 92A05

The Symposium was sponsored by the
National Science Foundation under Grant No. MPS-76-01249

Printed in the United States of America

CONTENTS

CONTENTS

FOREWORD

This volume contains lectures given at the Tenth Symposium on Some Mathematical Questions in Biology, held in Boston on February 24, 1976 in conjunction with the annual meeting of the American Association for the Advancement of Science. The Symposium was cosponsored by the American Mathematical Society and by the Society for Industrial and Applied Mathematics under the auspices of Section A, Mathematics, of the AAAS.

Three main themes are reflected in the lectures contained in this volume. The first two papers, by George Bell and Byron Goldstein, present a coordinated development of mathematical models in immunology. The mathematical foundations of this subject have been laid primarily within the past decade; and Bell, Goldstein, and their associates have played a major role. Their contributions provide a stimulating introduction to this vital new area of mathematical investigation.

Bell begins with a discussion of the biological fundamentals of the immune system, the kinetics of aggregation, and the theory of clonal selection by which the immune system learns to deal with unfamiliar antigens. He then discusses his own patching model, designed to explain critical phenomena in the activation of the cells which can secrete antibodies, and the mechanisms by which an individual avoids making antibodies to its own antigens. Goldstein follows with a survey of assay methods for circulating antibodies and for antibody-producing cells. He goes on to discuss two approaches for studying organism response to antigen injection, the clonal selection model of Bell and the optimality approach of Perelson, Mirmirani, and Oster. The complementarity of these two approaches parallels similar developments in evolutionary theory.

In the two middle papers in the volume, Hans Othmer and Stuart Kauffman

FOREWORD

address questions of significance for developmental biology. Kauffman's
paper is reprinted with minor change from D. L. Solomon and C. F. Walter (eds.)
Mathematical Models in Biological Discovery, Lecture Notes in Biomathematics,
Volume 13, Springer-Verlag, Heidelberg. Othmer, utilizing the basic diffusion-
reaction approach which has been shown to be of universal interest throughout
theoretical biology, surveys a variety of mechanisms which lead to spatial
pattern formation. Such questions arose a quarter-century ago in Turing's
discussion of morphogenesis, and continue to have major implications for
developmental biology. Kauffman examines the control of the periodic nature
of the cell cycle, investigating and contrasting a variety of models and
relating his investigations to other work on biological clocks. In particular
he presents his own limit cycle model, and assesses the current status of that
model in the light of experimental investigations.

The final two papers, by Sol Rubinow and Garrett Odell, address problems
in biomechanics, which have a rich history but have not been well represented
in this series. Rubinow discusses several fluid mechanical problems with
particular emphasis on blood flow in arteries and veins and on the swimming
of microorganisms. Odell develops a new continuum theory of axoplasmic
transport with special reference to the movement of nerve cell cytoplasm within
the confines of the axon membrane.

It is a pleasure, as in years past, to once again acknowledge the support
of the National Science Foundation and the staff of the American Mathematical
Society.

SIMON A. LEVIN

Lectures on Mathematics in the Life Sciences
Volume 9, 1977

SOME MATHEMATICAL MODELS IN IMMUNOLOGY, I.

George I. Bell[*]

1. INTRODUCTION

All higher animals employ a variety of mechanisms to pre-
serve their biological identity and integrity. Among the verte-
brates, the immune system is one of the most important. This is
a system of molecules (chiefly antibody or immunoglobulin molecules) and cells (chiefly lymphocytes) which serves to detect the
presence of foreign molecules (antigens) or cells and to hasten
their elimination from the body. In an adult human being there
are of the order of 10^{20} molecules and 10^{12} cells in the immune
system. The cells are continuously formed from division of more
primitive stem cells in bone marrow and are found in the circula-
tion and in organs of the lymphoid system, including the spleen,
thymus, and lymph nodes. The molecules are secreted by lymphocytes, circulate throughout the body, and, when bound to antigen,
interact with a variety of other cells and molecules so as to
effect the destruction or removal of the antigen.

The immune system has a number of properties which have combined to make its study, that is immunology, by far the most
highly developed field in mammalian biology. First of all, the
immune system is of great clinical importance, since it protects
us against multiplication of pathogenic bacteria, viruses, and

AMS(MOS) subject classifications (1970). Primary 92 A05,
9202; Secondary 05 C05

[*] Research performed under the auspices of the U.S. Energy
Research and Development Administration.

their toxic byproducts. There are however, a number of disease
states in which the immune system appears to be counter produc-
tive (e.g. rheumatoid arthritis) or ineffective (e.g. cancer); in
such cases it is hoped that improved understanding of the immune
system may facilitate therapy. Moreover, the immune system involves a singularly accessible and manipulable population of
cells; it has revealed many remarkable features concerning the
molecular biology of animal cells.

In addition, the immune system is a learning system, designe
to anticipate and cope with the unexpected. As a result of a firs
exposure to an antigen, an animal learns to respond more rapidly
and effectively to subsequent exposures. This learning is the
basis of vaccination and is explained by the clonal selection
theory, which we will describe in section III.

During the past fifteen years, this clonal selection theory
has become accepted as providing at least an outline of the functioning of the immune system and a number of quantitative techniques have been developed for measuring immune responses. However, many crucial details, particularly concerning initiation
and regulation of responses, remain poorly understood. Under
these circumstances, the mathematical analysis of problems in
immunology can be both stimulating to the mathematicians and
valuable to the immunologists.

2. THE MOLECULES

There are approximately twenty different classes of antibody
molecules, but their properties can be well understood by considering the members of a single class, namely the type G immunoglobulins, denoted IgG. IgG molecules are protein molecules having a molecular weight of around 150,000 and are composed of four
polypeptide chains.[*]

---

[*] There is a small amount of carbohydrate attached but its function is unknown.

For those who are unfamiliar with proteins I note that these
large biomolecules are made up of smaller molecular building
blocks called amino acids. Living organisms use twenty amino
acids in their proteins, though organic chemists can synthesize
lots more. The amino acids may be thought of as joined together
(by peptide bonds) in linear arrays which are called polypeptide
chains, though in fact, under the influence of intermolecular
forces, in solution these chains will usually arrange themselves
into fairly compact three dimensional structures. A typical protein contains several polypeptide chains which are joined by intermolecular bonds to form a three dimensional conformation precisely determined by the component amino acid sequences and the
solution properties.

Immunologists were long frustrated by the great heterogeneity of IgG molecules as they are found in, say, an animal's blood.
However by studying animals with cancer of antibody secreting
cells, they were able to isolate relatively homogeneous populations of antibody molecules and to determine their structure and
in many instances the amino acid sequences of the polypeptide
chains. It was then found that an IgG molecule contains two identical heavy chains having about 500 amino acids each and two identical light chains with about 220 amino acids. These are linked,
as sketched in Fig. 1, (by inter-cystein disulphide bonds) to
form a Y shaped molecule, which is so suggestive in its structure
as to represent the symbol of immunology. Each of the arms and
the stem are approximately 7 nm in length and 3.5 nm in diameter.

The amino acid sequences of antibodies are highly variable,
but, for antibodies of a single class, the variability is confined to one end of each chain. These variable ends are denoted
V_L and V_H and include about one-half and one-quarter of the amino
acids in the light and heavy chains respectively. Moreover, the
variability is concentrated in specific "hypervariable" regions
which are, in the folded molecule, near the binding site for

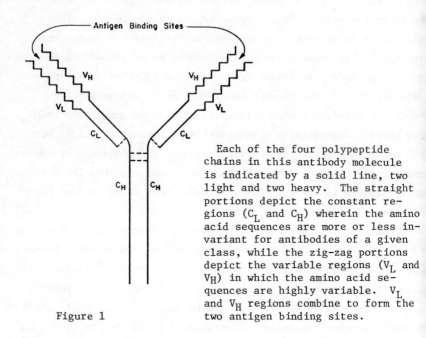

I₉G STRUCTURE

Figure 1

Each of the four polypeptide chains in this antibody molecule is indicated by a solid line, two light and two heavy. The straight portions depict the constant regions (C_L and C_H) wherein the amino acid sequences are more or less invariant for antibodies of a given class, while the zig-zag portions depict the variable regions (V_L and V_H) in which the amino acid sequences are highly variable. V_L and V_H regions combine to form the two antigen binding sites.

antigen. Since light and heavy V regions combine to form the binding sites, a great diversity of binding can be achieved with moderate diversity of V regions. With, for example, a repertoire of 10^3 V_L regions and 10^3 V_H regions it would be possible to construct antibody molecules having 10^6 different binding sites.

The constant regions of antibody molecules, and especially the heavy chain stem, determine the functioning of the antibodies once they are bound to antigen. For example, certain antibodies on binding to antigens cause the antigens to become appetizing to scavenger cells (macrophages) which will in turn ingest and frequently destroy the antigens. This function is determined by the constant region.

We thus see that the IgG molecule is beautifully designed
or diversity in antigen binding and constancy in function.   The
olecular mechanisms which make possible this design include
eparate DNA genes which code for V and C regions and the bring-
ng together (translocation) of these two genes for translation
nto a single polypeptide chain.

All antibody molecules have at least two identical antigen
inding sites.* Moreover most antigens have multiple identical
eterminants (epitopes) which can be bound by antibodies.   There-
ore antigen-antibody aggregates will tend to form whenever these
wo substances are present in solution together.   Many features
f this aggregation have been mathematically analyzed by making
he simplifying assumptions that the aggregates are topologically
rees and that all free sites of a given type (antigen or anti-
ody) are equivalent.   The first assumption neglects cycles with-
n an aggregate, as indicated in Fig. 2, while the latter neglects
ny difference, in reactivity or accessibility, between sites in
n aggregate and those on free molecules.

Immunologists have long recognized that the formation of
arge aggregates (precipitation) is impossible if either antigen
r antibodies are present in great excess in solution.   The rea-
ons are obvious: with antigen excess, most antibody molecules
ill be bound to two antigens and Ag-Ab-Ag complexes will pre-
ominate; with antibody excess, most antigen sites will be bound
o antibodies which are unlinked to other antigens.   Simple com-
inatorial calculations suffice for predicting the ranges of
antigen and antibody concentrations for which precipitation can
ccur.   Consider antigen molecules having f binding sites per
olecule and let P be the probability that a site is linked by

---

* Some antibody classes have more sites. For example, IgM molecules resemble five IgG molecules joined together in a flower-
like arrangement and have 10 sites.

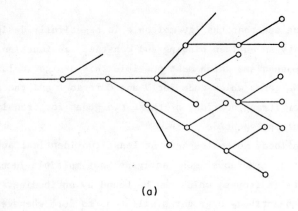

(a)

(b)    (b')

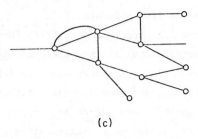

(c)

Figure 2. Aggregates. Lines represent antibody molecules with two
          binding sites per molecule. Circles represent antigen
          molecules with at least four sites for binding the anti-
          bodies. (a) A tree - breakage of any bond will lead to
          two disconnected smaller aggregates, or an aggregate
          plus a monomer. (b) Monogamous bivalent binding of one
          antibody molecule to one antigen molecule. When per-
          mitted, this complex may be much more stable than the
          combination of two antigen molecules to a single anti-
          body shown in (b'). (c) An aggregate with crosslinking:
          Some of the bonds can be broken without disconnecting
          the aggregate. The reactive partners in such bonds
          would be relatively likely to rejoin.

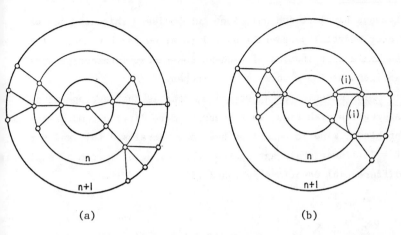

(a)   (b)

Figure 3.   Aggregates between bivalent antibodies, represented by
lines, and antigen molecules having at least four sites
per molecule. (a) A tree is arbitrarily centered and
only doubly bound antibody molecules are represented.
If the antigen molecules have f sites for binding
antibody, and P is the probability that a site is
connected to another antigen molecule, then the expected number of antigen molecules on the $(n+1)^{th}$ ring
is equal to the number on the nth ring times $(f-1)P$.
If $(f-1)P > 1$ there is a finite probability for the
aggregate to be of infinite size. (b) An aggregate
with intramolecular bonds. The bonds labeled (i) do
not serve to extend the aggregate but to stabilize it.

n antibody to another antigen molecule. Then if, and only if

$> 1/(f-1)$, there is a finite probability, [1,2] for an antigen

olecule to be in an aggregate of infinite size (in an infinite

ystem). This result is shown in Fig. 3. For the special case

f f = 2 , infinitely long linear chains will occur if P = 1 .

t is straightforward to relate P to the concentrations of

ntigen and antibody molecules and to their binding constants.

recipitation criteria have thereby been obtained [3-5] even when

he antibodies are heterogeneous, i.e., have a range of binding

onstants. The polymerization theory of Flory [1] and Stockmayer

6] has been used [7] to compute the equilibrium probability dis-

ribution of aggregate sizes when antigens interact with homo-

geneous bivalent and univalent antibodies. This approach was
later partially generalized [3] to allow for bivalent antibodie
having a distribution of binding constants, interacting with
antigens having a variety of epitopes.

The kinetics of aggregation, that is the development of
aggregates with time, is a somewhat more difficult mathematical
problem. Kinetic equations have been developed [8], and for
f = 2 (linear chain aggregates) the following system of nonlinear
differential equations is found [9].

$$\frac{dx_n}{dt} = 2k \sum_{m=1}^{n-1} x_{n-m}y_m - 2x_n(kS + k'n) + k' \sum_{m=n}^{\infty} (2x_m + y_m) ,$$

$$n = 1,2,\dots \qquad (1)$$

$$\frac{dy_n}{dt} = 4k \sum_{m=1}^{n} z_{n-m}x_m + k \sum_{m=1}^{n-1} y_{n-m}y_m - y_n\left[k(S+L) + (2n-1)k'\right]$$

$$+ 2k'\left[\sum_{m=n+1}^{\infty} x_m + \sum_{m=n+1}^{\infty} y_n + \sum_{m=n}^{\infty} z_m\right], \quad n = 1,2,\dots$$

$$(2)$$

$$\frac{dz_n}{dt} = 2k \sum_{m=1}^{n} z_{n-m}y_m - 2z_n(kL + k'n) + k' \sum_{m=n+1}^{\infty} (2z_m + y_m) ,$$

$$n = 0,1,2,\dots \qquad (3)$$

where $S = \displaystyle\sum_{m=1}^{\infty} y_m + 2 \sum_{m=0}^{\infty} z_m$ and $L = \displaystyle\sum_{m=1}^{\infty} (y_m + 2x_m)$, subject

o the initial conditions $x_1(0) = a$, $x_n(0) = 0$ (n=2,3,...),
$_n(0) = 0 = z_n(0)$ (n=1,2,...) , $z_0(0) = b$, with k and k'
eing non-negative constants. In these equations x_n, y_n, and
$_n$ are the concentrations of chains having n antigen molecules
lus n-1 , n, or n+1 antibody molecules respectively. Thus,
and L are the concentrations of free antibody and antigen
ites respectively while k and k' are the rates for bond for-
ation and breakage respectively. The individual terms in Eqs.
1)-(3) are simply interpretable. For example, the first term in
q. (1) represents the rate of formation of x_n due to reactions
f x_{n-m} with y_m . The equations have been solved by an indi-
ect combinatorial method [1,6,9] which is based on evaluating
he probability for each kind of aggregate, knowing the extent
o which antigen-antibody bonds have formed. The extent of re-
ction can, in turn, be readily calculated ignoring aggregation
ince free sites have been postulated to be equivalent whether
r not they are on aggregates.

For f > 2 , the non-linear differential equations describ-
ng growth of aggregates contain complicated statistical factors
hich reflect the multitude of ways in which large aggregates may
ecompose to form smaller ones. The required graph-theoretic
ounting problems do not appear to have been solved. Neverthe-
ess, the combinatorial method can still be used in order to ob-
ain the distribution of aggregate sizes [10].

When the simplifying assumptions (neglect of cyclic struc-
ures, equivalence of free sites) are relaxed, new effects are
ound to be important at least in certain cases. They have not
een analyzed in much generality.

Consider first the possibility that a single antibody mole-
ule may attach to two or more sites on a single antigen mole-
ule. Many biologically important antigens, such as viruses,
ave multiple identical epitopes. Moreover IgG molecules are
uite flexible in the "hinge region," that is the angle between

the two arms is variable. Therefore, attachment of both antibody
binding sites to a single multivalent antigen is often possible.
Both experiment [11] and analysis [12] have shown that, when
possible, such "monogamous-bivalent attachment" is much more
stable than single site attachment. The reason is that breaking
of a single antigen-antibody bond will lead to escape of a singly bound antibody but not of a doubly bound one. In the latter
case, the broken bond is likely to be restored, since the reacting partners are constrained to remain close together. Bivalent
attachment will evidently inhibit aggregation since each attachment removes two epitopes from a potential aggregate.

Consider next, cyclic structure within an aggregate, as
sketched in Fig. 3b. The simplest case is again of linear chains
(f=2) where the ends of a chain may join to form a circle.
Stockmayer [13] showed that for flexible chains and at low concentration, small circles are a preferred species at equilibrium.
More recently [14] the formation of knots during chain closure
has been considered, including both knots in a single closed
chain and the interlinking of different chains. Knotting is
likely only for relatively long linear chains which probably do
not occur in immunology, but it may be significant in other biological contexts such as the formation of circular DNA molecules.

For antigen molecules with more combining sites, a first
approach would be to retain the equivalent site hypothesis so
that every antigen site has some probability, P , of being connected to another site. In the presence of cyclic structure, the
critical value of P for precipitation must be larger since some
of the bonds as shown in Fig. 3b will not be useful in extending
the structure but serve to make redundant connections within
smaller aggregates [5]. If one assumes that the aggregates have
a regular geometrical structure, then the critical value of P
can be estimated from bond percolation theory [15,16] and is
found to exceed $1/(f-1)$ by 17% - 37% depending on the lattice

eometry [4]. Some polymerization experiments indicated a com-
arable increase in P , for f = 3 [1].

    Apart from uncertainties in the geometrical structure, per-
olation theory neglects an important distinction between bonds
n cyclic aggregates and those in trees. When any bond breaks
n a tree, the aggregate is thereupon divided into two smaller
ggregates which may diffuse apart. When a bond breaks in a cyc-
ic aggregate, the two reactive partners may still remain linked
hrough other bonds and are therefore relatively likely to become
ound again. Therefore cyclic structures will act to stabilize a
recipitate. The simplest example is the comparison of "monoga-
ous bivalent attachment" with attachment of a single antibody to
wo different antigen molecules. The cyclic complex, where per-
itted by spatial and orientational constraints, has a smaller
robability of dissociation per unit time by a factor $\sim 10^3$ [11,
2]. For similar reasons, small circles may be preferred rela-
ive to linear chains [13].

    I am not aware of analysis for more complex cyclic aggre-
ates, but it appears that the degree of bond stabilization will
ncrease with closeness and multiplicity of cyclic bonds, to
pproach a factor similar to that for monogamous bivalent attach-
ent. The argument is as follows. Consider an aggregate between
omogeneous antibodies and an antigen, having multiple copies of
 single epitope. The probability per unit time of breaking an
pitope-antibody bond is assumed to be the same for all the bonds
ut the probability of bond (re)formation is increased if the
artners are constrained to remain nearby. The tightest con-
traint is for monogamous-bivalent interaction where is has been
ssumed [12] that they are constrained to remain within a volume
etermined by the flexibility of the antibody hinge region. For
ycles of larger size, the constraints would generally be less
evere, unless multiple cyclic bonds served to enforce a rather
igid structure on the aggregate. Some related problems of

intramolecular reaction have been discussed by Morawetz [17].

Cycles presumably explain the virtual irreversibility of
many precipitates. That is, a tree-like precipitate should rapid
ly dissolve on resuspension in a solution lacking antibodies or
antigen; the time for dissolution would be the time for breaking
single bonds, typically $\sim$ one second. In fact however, most precipitates are far harder to dissolve.

Antigen-antibody complexes are important in clinical medicine since they are responsible for many of the symptoms, even
unto death, of disease. In addition, as we will see in the next
section, antigen-antibody aggregates form on the surfaces of cell
and may be required for initiation and regulation of an immune re
sponse. Finally, in the analysis of many immunological experiments, it is essential to distinguish between single site and
multivalent attachment of antibodies to antigens (or cells) and
to recognize possible effects of aggregation. Examples are given
in the following paper.

3. THE CELLS

According to modern versions of the clonal selection theory,
the cells of the immune system arise from division and differentiation of stem cells in the bone marrow. Those cells which are
potentially able to secrete antibodies are known as B-lymphocytes
or B-cells[*], and at some time during its development each B-cell
becomes committed to the expression of a single pair of V genes
and thereby, to the synthesis of a particular antibody molecule[†].

---

[*] B (bone marrow derived) lymphocytes are to be distinguished
from T-lymphocytes which must undergo further development in
the thymus. T-cells may help or suppress the activity of B-
cells or they may kill foreign cells, but they are not be-
lieved to secrete many antibodies.

[†] The descendents of a particular committed B-cell, which constitute a clone, continue to express the same V genes, there
by maintaining the antigen binding characteristics of all

The B-cell begins to insert these antibody molecules into its surface membrane where they function as receptors for recognition of complementary antigens. Under appropriate conditions, binding of complementary antigen to the receptors on a B-cell will activate that cell to proliferation and secretion of large amounts of antibody. A particular antigen may activate a hundred different clones of B-cells, each clone secreting its own antibodies which will bind the antigen. One result of this antigen-selective proliferation will be a large pool of "memory cells" which are rather similar to the original B-cells but are specific for the stimulating antigen. On subsequent exposure to the same antigen, these numerous memory cells will be able to more rapidly effect a response and thus clonal selection provides an explanation for the learning ability of the immune system. Moreover, as will be described in the following paper, there is a tendency for those cells which produce the best antibodies, i.e., those binding the antigen most tightly, to selectively proliferate. Therefore in a secondary response, high quality antibodies are produced from the start while in a primary response they take a longer time to develop.

The clonal selection theory was formulated by Jerne [20] and Burnet [21] and has become widely accepted as a valid outline of the operation of the immune system. There remain, however, many critical details concerning initiation and regulation of responses, which are poorly understood, controversial, and under active investigation in scores of laboratories throughout the world. In the remainder of this paper, I will consider two problems, in which mathematical analysis has been useful: (1) Under what circumstances will antigen binding activate B-cells? (2)

---

antibodies secreted by the clone. However later members of a clone may express different C genes and therefore secrete antibodies of different function [18]. There is some evidence for occasional alteration of V genes during antigen-driven proliferation [19].

How can the immune system distinguish self from non-self?

On the surface of a typical B-cell, there are approximately 10^5 receptors, each of which is an antibody molecule having two identical sites for binding antigen. These receptors are free to move along the cell membrane. Therefore, complementary multi-valent antigen molecules can cross-link receptors so as to form antigen-receptor aggregates on the surface of the B-cell. Such aggregates have been observed and are known as "patches." More-over, a metabolically active cell will sweep all of its patches to one pole to form a "cap" which may be then partically ingested and partially shed.

It appears that a sufficiently rapid rate of patch formation may be required for activation of B cells by molecular antigen[*]. I have therefore developed a model for patching kinetics [23,24] in which it is assumed that the first bond between an antigen molecule and a receptor is reversibly formed with forward and re-verse rate constants, k_a and k_d respectively, while subse-quent bonds are irreversibly formed with rate constant k_i , as sketched in Fig. 4. Note that patching is somewhat similar to precipitation discussed earlier, except that all the receptor molecules are bound to, but free to move along, a common surface. Thus, binding of one antigen molecule to more than one receptor will establish a cross-linked structure which is assumed stable in the model.

Let c be the concentration of free antigen, ρ and ρ_0 be the concentrations of free and total receptor sites, m be the concentration of singly bound antigen molecules, and M be the concentration of multiply bound antigen molecules, with on the average, f bonds per molecule. Then the equations of the

[*] Immunologists are far from being able to delineate a set of necessary and sufficient conditions for B-cell activation. In general, a proper cellular environment of the B-cell, including perhaps other cells which recognize the antigen, may be re-quired [22].

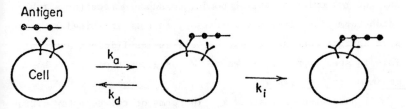

Figure 4.   Binding of multivalent antigen to cells. Multivalent antigen is assumed first to combine reversibly with a cell receptor, with forward and
reverse rate constants k_a and k_d. Then irrevers-
ible binding is established to further receptors
with rate constant k_i.

odel, depicted in Fig. 4, are

$$\rho(t) = \rho_0 - m(t) - fM(t) \tag{4}$$

$$\overset{\circ}{m}(t) = k_a c\rho(t) - k_d m(t) - k_i m(t)\rho(t) \tag{5}$$

$$\overset{\circ}{M}(t) = k_i m(t)\rho(t) \tag{6}$$

ι these equations $\overset{\circ}{M}$ is the proportional to the rate of patch
ormation. Solutions are sought for the initial conditions
$= \rho_0$, $M = m = 0$. After initial transients, it is reasonable,
ι the sense of singular perturbation theory, to set $\overset{\circ}{m} = 0$ in
ɿ. (5). In two interesting limits, simple equations for $\overset{\circ}{M}$ are
ɔund. If $k_d \ll k_i\rho$, $\overset{\circ}{M} = k_a c\rho$, while if $k_d \gg k_i\rho$,

$$\overset{\circ}{M} = k_i \frac{Kc}{(1 + Kc)^2} (\rho_0 - fM)^2 , \tag{7}$$

ιere K , the receptor-antigen equilibrium constant, equals
$_a/k_d$. In the former limit a singly bound antigen molecule is
ikely to become irreversibly bound and the rate of patching is
ιe rate of single bond formation. In the latter case, equilib-
ium is first established between free and singly bound antigen
ɔlecules; the lattice then gradually develops as singly bound

antigen molecules establish bonds, to unbound receptor sites.
Patch formation, like precipitation, is then inhibited by exces-
sive antigen concentrations, as well as insufficient antigen.
This is seen from Eq. (7) where $\overset{\circ}{M}$ as a function of c has a
maximum for $c = 1/K$.

I have also estimated k_i in terms of the receptor diffusion coefficient (along the membrane), D , and the mean spacing
between receptors, d , finding $k_i \propto fDd^{-2}$. Predictions of the
model have been compared with a variety of immunological data and
much agreement has been found [23,24]. One prediction is of particular interest. If a large value of $\overset{\circ}{M}$ is adopted as a necessary condition for B-cell activation, then a given antigen concentration, c , will preferentially select for activation those
cells having receptors with $K \simeq 1/c$. Evidence for such selection has been presented elsewhere [25,26] and will be further
discussed by Goldstein in the following paper.

The above model has been generalized by DeLisi and Perelson
[10] to take better account of the number of epitopes, f , per
antigen molecule. As a patch develops, they allowed the number
of bonds per antigen molecule to vary with time and thereby found
a more accurate dependence of patching rate on f . For example,
an antigen having $f = 2$ can only form linear arrays with receptors, so that patching is scarcely possible. Larger values of
f facilitate patch formation.

Patching is often induced by using antibodies against the
receptors (anti-immunoglobulin antibodies) instead of antigen as
the cross linking agents. While such antibodies, if IgG, have
only two binding sites per molecule they are not strictly equivalent to an antigen with $f = 2$. This is because the antibodies
are heterogeneous and hence more than two antibody molecules can
bind to the same receptor.

While the formation of receptor-antigen aggregates on B-cell
may be frequently associated with activation, it does not appear

be a sufficient condition, and may possibly lead to inactiva-
tion of the cell. There is evidence that at least some highly
multivalent antigens can directly activate B-cells, whereas antigens with fewer repeating determinants must be bound to a surface,
possibly a cell surface, before they can activate B-cells. In
general, it seems that the immune system uses more than simply
antigen binding for initiation and regulation of responses.

This brings us to the second key question: How does the
immune system distinguish self from non-self? Each individual
has many cells which would be recognized as antigenic and re-
jected by any other individual of the same species, except for an
identical twin*, and each individual has numerous molecules and
cells which are antigens to another species. So why doesn't an
individual make antibodies to his own antigens?

The information cannot be coded genetically since an individual will inherit (and express) antigenic determinants from
each parent and will inherit (but apparently not express) the
genetic information which is required to recognize these determinants as foreign.

The clearest distinction between self antigens and foreign
antigens is that, for the most part, the self antigens are always
present whereas the foreign antigens arise at particular times
after maturation of the immune system. Therefore the simplest
explanation for self-non-self discrimination [27] is that each
lymphocyte, as it matures, goes through a sensitive stage when
contact with the complementary antigen will be lethal. It then
needs to be arranged that each lymphocyte will encounter, with
high probability, any complementary self antigens during this
sensitive period. It has been observed that within the thymus of

---

Within some inbred strains, especially of mice, the individuals
are so genetically similar as to accept one another's cells.
They are effectively all identical twins.

a young animal there is extensive cellular proliferation but that
few of the resulting lymphocytes survive; it may be that self-
reactive lymphocytes are being eliminated in this process. Indeed, Jerne has suggested that self reactive T-cells would otherwise be abundant [28], but that during the proliferation and kill-
ing such cells are eliminated unless they by chance undergo a
mutation of a V gene so that they do not recognize self
antigens. The thymus would then be an organ for the generation,
by negative selection, of T-cells which recognize (even slightly)
different-from-self antigens. It may well be that elimination of
young lymphocytes is the basic mechanism* for self-non-self discrimination but two problems remain. The first is that immunologists have found various ways to paralyze or to kill mature B-
cells, for example by exposure to high doses of antigen. Assuming that non-reactivity to self antigens is brought about by such
paralysis, Bretscher and Cohn [29-31] have constructed an elegant
theory to explain how contact between antigen and B-cells will
sometimes lead to activation and sometimes to paralysis of the
cells. It is beyond the scope of this paper to describe this
model. It does explain some otherwise puzzling immunological experiments (including the requirement for T-cell - B-cell collaboration). It has not, for the most part, been cast into mathematical form, but it could and should be so formulated.

The second problem is that non-reactivity to self-antigens
has only been explained if the self-antigens are always present.
However some self-antigens are not always present, at least in
sufficient concentration to interest even a sensitive B-cell. In
particular, antigenic determinants are associated with the

---

*
Some immunologists believe that self reactive T cells are eliminated by such mechanisms and that B-cell activation requires
confirmation of foreignness by a nearby T cell so that self reactive B-cells although present cannot be activated. Such a
mechanism, like many others may fail on encountering antigens
which contain both self- and foreign-determinants, such as an
antigen-antibody complex.

ariable regions of antibody molecules. These <u>idiotypic</u> determi-
ants (or <u>idiotopes</u>) are more or less unique to each pair of V
egions and therefore each new clone that is selected for prolif-
ration by antigenic challenge, will in turn expose the immune
ystem to a new set of idiotypic antigens. Of course, prior to
hallenge one or some B-cells were present with these idiotopes
ut possibly in such low concentrations as to be unstimulating as
ntigens. Moreover it is likely that entirely new idiotopes [32]
rise during development and proliferation of lymphocytes, both
rior to and perhaps following [19] exposure to antigen.

There is no doubt that individuals can recognize the idio-
opes on their own antibodies as antigens and that they can be
nduced to make antibodies (anti-idiotypic antibodies) against
diotopes [33,34]. Anti-idiotypic antibodies might be viewed as
ather undesirable and generally unimportant by-products of the
mmune system but it has been recently argued that, quite to the
ontrary, they may be critical for normal regulation of immune
esponses [33].

According to this latter view, the immune system, in the ab-
ence of antigen stimulation, should be regarded as a dynamic net-
ork of V-regions, which are displayed both on cell surfaces and
n free molecules, and are interacting by means of idiotope recog-
ition.[*] Such a network can be enormously complex, comparable in
omplexity perhaps to the nervous system, not only because of the
arge number of potential idiotope-paratope pairs, but also be-
ause the interactions can be either stimulating or repressive to
ne or both of the interacting partners. Consider, for example,
lymphocyte having receptors with idiotopes/paratopes which are

---

Antigenic determinants associated with unique V regions are
called idiotopes. There may be several per antibody V region.
The site which binds to an antigen is called a <u>paratope</u>. There
is, by definition, but one paratope per $V_L \times V_H$ region. Paratopes recognize and bind idiotopes. Idiotopes may be thought
of as bumps and paratopes as clefts on V regions.

bound to paratopes/idiotopes of other antibody molecules (which
may be attached to another cell). Insofar as the lymphocyte is
interacting with an antigen it may be stimulated but insofar as
it is interacting with antibodies, it may be killed. The actual
outcome of the interaction may depend on the type of lymphocyte
(T or B-cell, maturity), class of antibody (whether it is appetizing to macrophages, binds complement, is attached to cells .
..), whether the lymphocyte has the idiotope or paratope, etc.

Three simplified mathematical models have been devised for
modeling idiotope networks [35-37]. They include both excitatory
and inhibitory interactions so that the immune system can exist
in several steady states and can be switched from one to another
by exposure to antigens.

The development and testing of idiotope network models is an
exciting possible enterprise for theorists in immunology. But
the crucial question of whether network interactions play any important role in normal immune responses remains to be answered.

Mathematicians who seek further exposure to the facts of
immunology may consult one of the good recent textbooks [e.g.,
38,39]. Beyond that I strongly support discussions and possibly
collaboration with immunologists.

REFERENCES

1.   P. J. Flory, "Molecular Size Distributions in Three Dimen-
     sional Polymers," J. Am. Chem. Soc. 63 (1941) 3083-3100.

2.   M. Gordon, "Good's Theory of Cascade Processes Applied to
     the Statistics of Polymer Distributions," Proc. Roy. Soc.
     A. 268 (1962) 240-256.

3.   F. Aladjem and M. T. Palmiter, "The Antigen-Antibody Reac-
     tion, V. A Quantitative Theory of Antigen-Antibody Reactions
     which Allows for Heterogeneity of Antibodies," J. Theoret.
     Biol. 9 (1965) 8-21.

4.   G. I. Bell, "Mathematical Model of Clonal Selection and
     Antibody Production, II.," J. Theoret. Biol. 33 (1971) 339-
     378.

. C. DeLisi, "A Theory of Precipitation and Agglutination Reactions in Immunological Systems," J. Theoret. Biol. $\underline{45}$
(1974) 555-575.

. W. H. Stockmayer, "Theory of Molecular Size Distributions
and Gel Formation in Branched-Chain Polymers," J. Chem. Phys.
$\underline{11}$ (1943) 45-55.

. R. J. Goldberg, "A Theory of Antibody-Antigen Reactions. I.
Theory for Reactions of Multivalent Antigen with Bivalent
and Univalent Antibody," J. Am. Chem. Soc. $\underline{74}$ (1952) 5715-
5725.

. A. S. Perelson and C. DeLisi, "A Systematic and Graphical
Method for Generating the Kinetic Equations Governing the
Growth of Aggregates," J. Chem. Phys. $\underline{62}$ (1975) 4053-4061.

. A. S. Perelson and C. DeLisi, "Infinite Systems of Coupled
Quadratic Non-Linear Differential Equations," SIAM Review,
(in press).

. C. DeLisi and A. Perelson, "The Kinetics of Aggregation
Phenomena, I. Minimal Models for Patch Formation on Lymphocyte Membranes," J. Theoret. Biol. (in press).

. C. L. Hornick and F. Karush, "Antibody Affinity - III. The
Role of Multivalence," Immunochem. $\underline{9}$ (1972) 325-340.

. D. M. Crothers and H. Metzger, "The Influence of Polyvalency
on the Binding Properties of Antibodies," Immunochem. $\underline{9}$
(1972) 341-357.

. H. Jacobson and W. H. Stockmayer, "Intramolecular Reaction
in Polycondensation. I. The Theory of Linear Systems," J.
Chem. Phys. $\underline{18}$ (1950) 1600-1606.

. M. D. Frank-Kamenetskii, A. V. Lukashin and A. V. Vologodskii,
"Statistical Mechanics and Topology of Polymer Chains,"
Nature $\underline{258}$ (1975) 398-402.

. H. L. Frisch and J. M. Hammersley, "Percolation Processes
and Related Topics," J. Soc. Indust. Appl. Math. $\underline{11}$ (1963)
894-918.

. J. W. Essam, "Percolation and Cluster Size," in Phase Tran-
sitions and Critical Phenomena, $\underline{2}$, ed. C.Domb and M. S.
Green, Academic Press (1972) 197-270.

17.   H. Morawetz, "Kinetics of Intramolecular and Intermolecular
      Reactions Involving Two Functional Groups Attached to Poly-
      mers," Pure and Appl. Chem. 38 (1974) 267-277.

18.   P. J. Gearhart, N. H. Sigal and N. R. Klinman, "Production
      of Antibodies of Identical Idiotype but Diverse Immunoglob-
      ulin Classes by Cells Derived From a Single Stimulated B
      Cell," Proc. Nat. Acad. Sci. USA 72 (1975) 1707-1711.

19.   A. J. Cunningham and S. A. Fordham, "Antibody Cell Daughter
      can Produce Antibody of Different Specificities," Nature
      250 (1974) 669-670.

20.   N. K. Jerne, "The Natural Selection Theory of Antibody For-
      mation," Proc. Nat. Acad. Sci. USA 41 (1955) 849-857.

21.   F. M. Burnet, The Clonal Selection Theory of Acquired Immu-
      nity, Vanderbilt University Press, Nashville, Tennessee
      (1959).

22.   Transplantation Reviews, Vol 23, "Concepts of B lymphocyte
      Activation," ed. G. Möller (1975).

23.   G. I. Bell, "Model for the Binding of Multivalent Antigen
      to Cells," Nature 248 (1974) 430-431.

24.   G. I. Bell, "B Lymphocyte Activation and Lattice Formation,
      Transplant. Rev. 23 (1975) 23-36.

25.   G. W. Siskind and B. Benacerraf, "Cell Selection by Antigen
      in the Immune Response," Adv. Immunology 10 (1969) 1-50.

26.   G. I. Bell, "Mathematical Model of Clonal Selection and
      Antibody Production," J. Theoret. Biol. 29 (1970) 191-232.

27.   J. Lederberg, "Genes and Antibodies," Science 129 (1959)
      1649-1653.

28.   N. K. Jerne, "The Somatic Generation of Immune Recognition,
      Eur. J. Immunol. 1 (1971) 1-9.

29.   P. Bretscher and M. Cohn, "A Theory of Self-Nonself Discrim-
      ination," Science 169 (1970) 1042-1049.

30.   P. Bretscher, "The Control of Humoral and Associative Anti-
      body Synthesis," Transplant. Rev. 11 (1972) 217-267.

31.   P. Bretscher, "On the Control Between Cell-Mediated, IgM,
      and IgG Immunity," Cell. Immunol. 13 (1974) 171-195.

32.   M. Cohn, "The Molecular Biology of Expectation," in <u>Nucleic Acids in Immunology</u>, eds. O. J. Prescia and W. Braun, Springer Verlag: New York (1968).

33.   N. K. Jerne, "The Immune System – A Web of V-Domains," The Harvey Lectures presented March 1975 (to be published by Academic Press).

34.   N. Sakato and H. N. Eisen, "Antibodies to Idiotypes of Isologous Immunoglobulins," J. Exp. Med. <u>141</u> (1975) 1411-1426.

35.   P. H. Richter, "A Network Theory of the Immune System," Eur. J. Immunol. (to be published).

36.   G. W. Hoffmann, "A Theory of Regulation and Self-Nonself Discrimination in an Immune Network," Eur. J. Immunol. <u>5</u> (1975) 638-647.

37.   G. Adam and E. Weiler, "Theory of Ontogenetic Generation and Subsequent Limitation of Antibody Diversity by Somatic Mutation and Anti-Idiotypic Elimination," Eur. J. Immunol. (to be published).

38.   I. M. Roitt, <u>Essential Immunology</u>, 2nd Edition, Blackwell Scientific Publishing, London (1974).

39.   B. D. Davis, R. Dulbecco, H. Eisen, H. Ginsberg and W. B. Wood, <u>Microbiology</u>, 2nd Edition, Hoeber Med. Div. – Harper & Row, New York (1974).

THEORETICAL DIVISION
UNIVERSITY OF CALIFORNIA
LOS ALAMOS SCIENTIFIC LABORATORY
LOS ALAMOS, NEW MEXICO 87545

ectures on Mathematics in the Life Sciences
olume 9, 1977

SOME MATHEMATICAL MODELS IN IMMUNOLOGY, II.
Byron Goldstein[*]

1. INTRODUCTION

A minimum requirement for a mathematical model of a biolog-
cal system is that it exhibit those properties of the system
hat have been well established through experimental investiga-
ion. In many experiments in immunology the quantities of inter-
st cannot be directly measured and therefore the interpretation
f these experiments is not straightforward. Establishing the
ystem characteristics even after much experimental data has been
ccumulated is not a simple task. Mathematicians can make use-
ul contributions to the field of immunology in two ways, by aid-
ng in the analysis and design of experiments and by mathemati-
ally modeling various aspects of the immune system. To illus-
rate I will first look at how immunologists have gone about an-
wering the question, "What happens to an animal after it is in-
ected with an antigen?", and then look at how some mathematical
odelers have attempted to answer the same question.

In general, if a vertebrate is exposed to an antigen two
ypes of immune responses can occur. The mediators of the re-
ponses are either sensitized lymphocytes (cell mediated response),
r specifically reacting antibody molecules (humoral response).

AMS(MOS) subject classifications (1970). Primary 92 A05,
202: Secondary 05 C05.
    [*] Research performed under the auspices of the U.S. Energy
esearch and Development Administration.

Immunity to tubercle bacilli, rejection of skin transplants from genetically different individuals of the same species and poison ivy are all examples of cell-mediated responses, while the induction of immunity to such diseases as diphtheria, tetanus, smallpox and measles through vaccination are examples of humoral responses. In this paper only the humoral response is discussed.

When an animal is inoculated with an antigen which it has not previously been exposed to, it mounts a primary immune response. In general detectable concentrations of antibodies which bind specifically to the antigen appear in the serum between one and thirty days after inoculation. With time the concentration rises to some maximum value and then gradually drops off.[*] The first antibodies to be detected are of the IgM class (in most species IgM molecules are composed of five subunits, each of two heavy and two light chains, the total molecule having ten binding sites) followed shortly after by the detection of IgG. As the response progresses the serum concentration of IgM drops off much more rapidly than that of IgG; in the latter stages of the response only IgG is detectable.

When an animal is exposed to the same antigen some time after the primary response, the new response, the secondary response, is considerably enhanced. It is characterized by a much shorter time between exposure to antigen and appearance of antibodies, a faster initial rise in antibody concentration, the attainment of much higher maximum concentrations, and the persistence of detectable concentrations for considerably longer periods. Further, lower initial doses than are required to elicit the primary response, will cause a secondary response. The antibodies produced in the secondary response are overwhelming IgG. The ability to produce an enhanced secondary response,

---

[*]   Although this is what is usually observed, a cycling of the antibody concentration with time after a single injection of antigen has been reported [1].

hich is commonly referred to as immunological memory, can last
or many years and is what provides one with long lasting immu-
ity against many infectious diseases.

II.   THE ASSAYS

.   Circulating Antibodies

It was pointed out in the previous paper that the cells
hich possess the potential to produce antibodies are the B-
ymphocytes or B-cells and that a particular antigen may stimu-
ate a number of different B-cells to proliferate and secrete
ntibodies. B-cells from different clones will secrete anti-
odies which differ in their variable region and these differ-
nces will be predominantly in the sequences of amino acids
hich make up the antibody binding sites. In the serum of an
nimal responding to an antigen challenge, the multiclone re-
ponse is reflected in the heterogeneous binding characteristics
f the antibodies.*

In order to determine the binding characteristics of an
ntiserum (a serum containing antibodies against a specific anti-
en) immunologists have made wide use of relatively simple anti-
ens. Indeed, much of what is know about the binding properties
f antibodies has been gained through the use of small molecules,
alled haptens, which are not immunogenic (not capable of stimu-
ating antibody production), but become immunogenic when coupled
o proteins (when a hapten-protein complex is injected into an
nimal, antibodies which bind specifically to the hapten appear
n the serum).

The binding of a monovalent hapten to a single antibody site
s characterized by two parameters the forward and reverse rate
onstants k_a and k_d. At equilibrium, for a homogeneous anti-

---

* There are some antigens which produce a monoclonal, or at least
very restricted response, since the serum's binding character-
stics appear homogeneous [2].

serum, the number of haptens bound per antibody molecule is given
by the equation

$$r = nKH/(1 + KH) \qquad (1)$$

where n is the valence of the antibody (2 for IgG, 10 for IgM),
H is the free hapten concentration and K is the hapten-anti-
body binding site affinity, i.e., $K = k_a/k_d$.

For a heterogeneous antiserum there is a distribution of
affinities and

$$r = nH \int_0^\infty Kf(K)dK/(1 + KH) \qquad (2)$$

where $f(K)$ is the normalized affinity distribution.[*]

Considerable effort has gone into determining $f(K)$ since
it contains information about the biological processes involved
in antibody production and removal. Attempts to directly determine $f(K)$ by fractionation of serum on the basis of affinity
through the use of immunoabsorption techniques, although useful
in revealing general properties, have so far not yielded an accurate picture of the affinity distribution. The fractions obtained tend themselves to be heterogeneous with respect to affinity and overlap in their spread of K values [4].

Most of the information known about $f(K)$ has been obtained
indirectly by measuring the fraction of sites bound as a
function of the free hapten concentration and then attempting to

---

[*] Since the number of clones involved in an immune response is
finite the affinity distribution is discrete. Equation (2) is
written as if $f(K)$ is continuous simply for convenience. In
some cases it is reasonable to treat the distribution as continuous since the number of clones involved in the response is high.
For example, for the hapten n-nitro-p-idophenyl (NIP) it has been
estimated that from 8,000-15,000 clones in a single mouse can
form anti-NIP [3].

ess   the distribution that will fit the data or use the data

"unfold" Eq. (2) and obtain f(K).

In most cases experimentalists have been content to deter-
ne two parameters from the binding data; one a measure of the
erage affinity, the other a measure of the width of the dis-
ibution. Although it is possible to experimentally determine
e first two moments of the affinity distribution [5], and there-
re the average affinity and the variance, this is in general
t done. Instead a particular distribution is assumed and the
rameters which describe the distribution are determined.

One drawback of this procedure is that when the assumed dis-
ibution is not a good approximation to the experimental distri-
tion, it is not clear how the determined model dependent param-
ers are related to the average affinity and variance of the
perimental distribution.

Another problem is that the values of the experimentally
termined parameters depend to some extent on the assay used to
asure r. (For a discussion of the various assays see chapter
of the text by Eisen [6] as well as [18].) The classic method
r measuring r is that of equilibrium dialysis. The method
asures the free and bound hapten concentrations unambiguously,
it has its limitations; it is time consuming and it cannot
used to study nondialyzable antigens. Another method fre-
ently used, which is considerably faster, is the Farr assay [7].
is based on the observation that antibodies precipitate in 50%
turated ammonium sulfate. In such a solution the free hapten,
ich will not precipitate, can be separated from the bound hap-
, which will precipitate. Studies which measure the "average"
finity* of the same antiserum using both methods tend to find
litative but not quantitative agreement [8]. There is still

---

average appears in quotation marks because the "average" affin-
es obtained are weighted averages, but how they are weighted
many cases is not clear. They are rarely the first moment of
distribution.

a need to clarify how the parameters determined by various immunological assays are related, and how they depend on the affinity
distribution of the antiserum.[*]

Despite the difficulties just discussed, some very important
properties of the immune response have been established from determining "average" affinities as a function of antigen dose,
time after immunization, type of antigen, etc. The most striking
finding, known as the maturation of the immune response, is that
the "average" affinity progressively increases with time after
immunization [10-12]. Further, the "average" affinity increases
more rapidly with lower doses of antigen. No clear trend in the
behavior of the width of the distribution as a function of time
has been established.

In the patching model of Bell's [13] (discussed in the previous paper) which assumes that a sufficiently rapid rate of
patch formation will activate B-cells, maturation of the affinity
follows in the limit that equilibrium is established between free
antigen molecules and antigen molecules bound to B-cell surface
receptors, i.e., the limit where an antigen molecule goes on and
off a B-cell many times before it is involved in a cross-linking
step. Bell found that cells with $K \sim 1/H$, where H is the
antigen concentration will be preferentially stimulated.[†] As
the antigen concentration decreases, clones which produce higher
affinity antibodies will be selected for; this will be reflected
in the serum as an increase in the affinity.

There has been one series of studies where affinity distributions were obtained [15-18]. Equilibrium dialysis measurements were made to obtain r as a function of the free hapten

---

[*] A quantity which is sometimes measured by the Farr assay is the
antigen binding capacity (ABC). A theoretical study of the dependence of ABC on the affinity distribution and the antibody
concentration has been carried out [9].

[†] This condition is preserved in the detailed model of patch
formation recently proposed by DeLisi and Perelson [14].

concentration.   The unfolding was accomplished by approximating
the true distribution by a number of subpopulations, i.e.,

$$\frac{r}{n} = \frac{K_1 H N_1}{1 + K_1 H} + \frac{K_2 H N_2}{1 + K_2 H} \cdots + \frac{K_n H N_n}{1 + K_n H} \tag{3}$$

where N_i is the fraction of antibodies with affinity K_i and
H is the free hapten concentration.

The set of affinities and the number of subpopulations were
chosen first and then, using a computer, a curve fitting procedure was employed to determine the set of N's. Since both the
true number of subpopulations and the affinities associated with
them are unknown a definitive distribution cannot be computed.
(In their most careful studies 12 subpopulations were used with
the affinity values ranging from $10^3 - 10^{11} M^{-1}$.)

To test their unfolding procedure, known distributions were
used to generate binding "data;" then the binding "data" was used
to obtain the distributions. Shown in Fig. 1 are the results

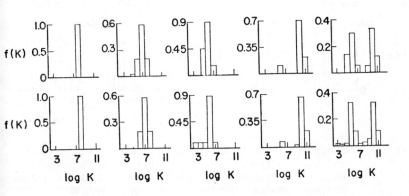

Figure 1. In the first row are five hypothetical affinity distributions f(K) that were used to generate binding date.
In the second row are the distributions that were obtained from this data by Weblin and Siskind using their
unfolding procedure. K is the affinity. [T. P. Weblin
and G. W. Siskind, Immunochem. 9 (1972) 987-1011].

obtained for five distributions. Although some of the details are not recaptured there is good overall agreement.

This procedure was used to study the immune response in rabbits to a hapten-protein carrier.[*] There was considerable variation from rabbit to rabbit but the following generalizations can be made: Initially the antiserum is composed of a heterogeneous distribution of low affinity antibody. As time progresses higher affinity antibodies appear and gradually come to make up a major portion of the distribution. For very long times, of the order of a year, the percentage of high affinity antibodies drops and the average affinity decreases. This latter observation does not mean that cells capable of producing higher affinity antibody are no longer present in the B-cell population since, if a second challenge with the same hapten-protein carrier is performed, high affinity antibodies very quickly appear in the serum. In Fig. 2 one such study is shown.

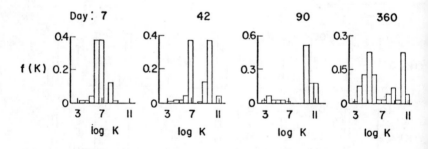

Figure 2. Shown are affinity distributions calculated by Weblin and Siskind from binding data obtained from measurements on the serum of a rabbit immunized with a hapten-protein complex (DNP-BGG). The rabbit was bled at various times after immunization (7, 42, 90 and 360 days). [T. P. Weblin and G. W. Siskind, Immunochem. 9 (1972) 987-1011].

---

[*] The hapten-protein carrier was dinitrophenylated bovine gamma globulin (DNP-BGG).

The procedure just outlined, although it gives an approximate picture of the distribution, is not completely satisfying.
Recently there has been an attempt to obtain an analytic form for
the distribution [19] based on the fact that when Eq. (2) is re-
written as follows,

$$r(1/Y) = n\int_0^\infty \frac{Kf(K)dK}{K + Y} \tag{4}$$

where $Y = 1/H$, $r(1/Y)$ is the Stieltjes transform of $nKf(K)$.
The problem then is to compute the inverse Stieltjes transform
of $r(1/Y)$. The solution of this is well known.* The real difficulty is to arrive at an analytic expression for r from the
experimental data points. Bruni and his colleagues [19] have
obtained an analytic expression for r that depends on four
parameters. The analytic form they have chosen yields a bimodal
distribution (two peaks), which for some data gives good agreement (see Fig. 2). However, the analytic form for r used by
Bruni can lead to predictions of two peaks, when in fact there is
only one present. There is still work to be done on the unfolding problem.

Antibody Producing Cells

Although considerable insight into the underlying cellular
dynamics of the humoral response has been gained by studying the
binding properties of antisera, a detailed understanding requires
measurements at the cellular level. The most widely used technique for studying single antibody forming cells (AFC) is the
Jerne hemolytic plaque assay [20]. Since its introduction, when
it was used simply to detect AFC, the technique has undergone

---

If for each K', for $0 \leqslant K' < \infty$, the integral $\int_0^{K'} |Kf(K)|dK$
and the integral in Eq. (4) is finite, the solution to Eq. (4) is
[19]:
$$f(K) = (1/\pi K_n)\,\text{Im}[r(e^{i\pi}/K)].$$

many modifications to both improve its sensitivity and extend its
usefulness [21].

In a typical experiment to detect AFC a mouse is immunized
with sheep red blood cells (RBC). Some time later the spleen of
the mouse is removed, cells from the spleen are mixed with the
sheep RBC, and the mixture is plated in a thin layer of agar.[*]
The plate is incubated, usually for one hour, and complement is
added. Complement is a group of proteins which occur naturally
in serum and causes the lysis of any RBC which has a sufficient
number of antibodies bound to its surface.[†] What is observed are
clear areas of lysed RBC called plaques, with each plaque having
at its center a lymphocyte. In order for a plaque to be produced
the lymphocyte at the center of the plaque must be an AFC.

The method is not restricted to using only red blood cell
antigens. For example, there are a variety of techniques for
covalently coupling haptens to red blood cell surfaces [21]. By
using haptenated red blood cells in the layer containing AFC, the
Jerne plaque technique has been successfully employed to study
the response of various animals to hapten-protein complexes.

The time development of plaques has been studied by initially incorporating complement in the layer along with the sheep
red blood cells and spleen cells. It was found that for many
plaques produced by IgM the plaque radius, R, increased with
time in the following way [24-26].

---

[*] Liquid layer techniques which do not use agar are also widely
employed [22].

[†] Complement can lyse a RBC with a single IgM molecule bound
to its surface. However, it takes two IgG molecules anchored to
the membrane in close proximity before complement can act.
Humphrey has estimated that approximately 800 IgG molecules must
be bound to a sheep red blood cell surface before it will lyse
in the presence of complement [23].

$$R^2 = k(T - T_0) \tag{5}$$

here T is the total time of the experiment and k and T_0
re constants for a given plaque. T_0 is presumably the time at
iich the AFC began emitting antibodies. However, measurements
or times less than a few minutes cannot be made since the experi-
entalist does not know where in the layer a plaque will start.
e must first scan the layer until he find a plaque, and then
ollow it. T_0 is obtained by extrapolating the plot of R^2 vs
  back to $R^2 = 0$.

A considerable effort has gone into the mathematical descrip-
ion of various plaque experiments. All the work, with one excep-
ion [21], is based on the diffusion-reaction equations written
own by DeLisi and Bell to describe the growth of plaques [27].
o see how these equations arise, consider an AFC in a thin layer
f thickness h , which emits IgM antibodies at a rate S(t). In
he layer the antibodies diffuse with a constant diffusion coef-
icient D , and bind reversibly to various sites on the RBC sur-
aces. To model the binding of IgM I will greatly simplify the
escription by using only two rate constants, k_1 , and overall
orward rate constant, and k_2 , an overall reverse rate constant.
he equations then for the free antibody concentration, c , and
he bound site concentration ρ are:

$$\frac{\partial c}{\partial t} = D\nabla^2 c - \frac{\partial \rho}{\partial t} + \frac{S(t)}{h}\,\delta(\underset{\sim}{r}) \tag{6}$$

$$\frac{\partial \rho}{\partial t} = k_1 c(\rho_0 - \rho) - k_2 \rho \tag{7}$$

here ρ_0 is the initial concentration of free binding sites,
.e., $\rho_0 = e\rho_{RBC}$, where e is the number of binding sites per
BC (the epitope density), and ρ_{RBC} is the RBC concentration
n the layer. The AFC has been taken to be at the origin; $\delta(\underset{\sim}{r})$
s the two dimensional Dirac delta function.

Equation (6) states that the free antibody concentration changes by diffusion, by binding to, or being released from, RBC surfaces, and, at the origin, by the production of antibodies at a rate $S(t)$. Equation (7) states that bound sites are formed when antibodies bind to RBC surfaces; the rate is proportional to the product of the free antibody concentration and the free site concentration. Bound sites are lost when antibodies come off the RBC surfaces; the rate of release is proportional to the concentration of bound antibodies. (For a more complete discussion of these equations see reference [28].)

IgM, since it has a valence of ten, is capable of forming long lasting multisite attachments to RBC surfaces providing e, the density of binding sites, is high enough. Simple estimates, which assume the sites are uniformly distributed on the RBC surface, indicate that if $e < 1.1 \times 10^5$ sites/RBC, multisite attachment is unlikely [30]. Estimates of e for various red blood cell surface antigens range from 10^3-10^6 sites/RBC (see p. 174 of reference [21].) For haptenated red blood cells the range is similar. It is reasonable to assume that both short-lived single site attachments and long-lived multisite attachments are physically realizeable. The solutions to Eqs. (6) and (7) have been studied for both types of binding [27,29,30]. I will discuss in detail only the single site attachment case.

It is the linearized versions of Eqs. (6) and (7) that have been solved. The linearization is achieved by assuming that the fraction of sites bound on a single red blood cell is small and hence $\rho_0 - \rho \approx \rho_0$. This is expected because the free antibody concentrations which develop in a layer from the emission of antibodies by a single AFC, at least for times less than many hours, are too small to cause large fractions of sites to be bound. Of course, after the solutions are obtained one can calculate the fraction bound at any r and t and see if the solutions are consistant with this assumption.

Rate constants for single site attachment of antibodies are known [31]. IgM forms very weak single site attachments, which rapidly dissociate ($k_2^{-1} \lesssim 10$ sec). This implies that when only single site attachment can occur, there is local equilibrium between the free and bound antibody concentrations. The local equilibrium assumptions can be written as follows:[*]

$$\rho = K\rho_0 c \ . \tag{8}$$

With the local equilibrium assumption and the assumption that the fraction of sites bound is small, Eq. (7) becomes simply a diffusion equation,

$$- \frac{S(t)}{hD} \delta(\underset{\sim}{r}) = \nabla^2 c - \frac{1}{D^*} \frac{\partial c}{\partial t} \tag{9}$$

where $D^* = D/(1 + K\rho_0)$. If the AFC emits at a constant rate,
$$(t) = \begin{cases} 0 & t < 0 \\ S & t \geqslant 0 \end{cases} \ , \text{ where the duration time } t = T - T_0 \ , \text{ then}$$
the solution to Eq. (9) is

$$c(r,t) = \frac{S}{4\pi Dh} E_1(r^2/4D^*t) \tag{10}$$

where $E_1(\cdot)$ is the exponential integral of order one.[†]

What one wants is an equation for the plaque radius, not the free antibody concentration. To obtain such an equation I assume that at the plaque radius there are always a fixed number N of antibodies bound per RBC; N is of the order of 1 for IgM. At

---

At equilibrium $\rho = K\rho_0 c/(1 + Kc)$, but when the fraction of sites bound is small, $Kc \ll 1$ and $\rho \simeq K\rho_0 c$.

This is the solution for diffusion in a plane which is appropriate for the thin layers usually used. It is straight forward to solve the finite layer problem by the method of images.

the plaque radius $\rho = N\rho_{RBC}$ and from Eq. (8) and (10) it follows that

$$\frac{4\pi DhN}{KeS} = E_1(R^2/4D*t) \tag{11}$$

where R is the plaque radius.

Since the left hand side of Eq. (11) is a constant for a particular AFC the argument of E_1 must also be a constant, i.e.,

$$R^2/4D*t = \text{const.} \tag{12}$$

This is just the behavior that has been observed for direct plaques [see Eq. (5)]. To date, studies comparing the time development of plaques when densely or sparsely haptenated red blood cells are used have not been done. The theory predicts that if densities can be reached at which IgM forms multisite attachments, the time development of plaques will deviate from the $R^2/t = $ constant behavior.

I would now like to discuss a technique, called plaque inhibition, which was introduced to determine antibody affinity at the level of single cells [32]. Such a technique is essential in studying the maturation of the immune response. In plaque inhibition experiments haptenated RBC, spleen cells and free hapten are present in the same layer. The free hapten competes for the antibody binding sites with the hapten that are coupled to the RBC surfaces. From a single spleen a number of plates are made up, each having different concentrations of free hapten. A plot, called a plaque inhibition curve, similar to the one in Fig. 3, is made of the number of plaques, N_p , vs the hapten concentration, H . N_p goes to zero for large H because at large hapten concentrations the antibody binding sites are taken up with free hapten and therefore cannot bind to the hapten on the red blood cell surfaces.

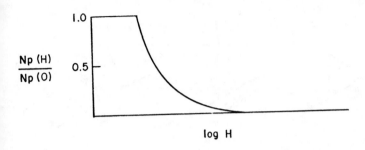

Figure 3. A typical hapten inhibition curve. $Np(H)$ is the number of plaques observed when the layer contains free hapten at concentration H. $Np(0)$ is the number of plaques when there is no free hapten present.

The plaque inhibition curve is used to infer information about the distribution of antibody affinities at the cellular level. In particular, $1/H_{50}$, where H_{50} is the hapten concentration needed to inhibit 50% of the plaques is taken as a measure of the average affinity. The qualitative argument which immunologists have used to justify their conclusions is as follows: high affinity antibodies will bind hapten more readily than low affinity antibodies; thus plaques produced by high affinity antibodies will be inhibited by less free hapen than plaques produced by low affinity antibodies [33].

    Considerable work has been done on the mathematical theory of plaque inhibition [34-37]. In the simplest case where all the AFC emit at a constant rate for the entire experiment, there are three distributions which must be considered: the distribution of affinities, of emission rates and of initial starting times. There is little experimental information on the last distribution. In the few plaque time development studies a lag has been observed of from 0-15 minutes before antibody emission begins [24-26]. For lack of information it has been assumed that although AFC do not all start to emit antibodies at the same time, the

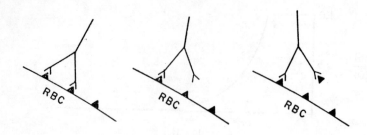

-Figure 4.   The three types of attachments on antibody can undergo
             at a red blood cell (RBC) surface. ▲ represents both
             the hapten coupled to the RBC surface and the free
             hapten.

distribution in starting times is narrow and can be ignored.

The theory is considerably easier for IgG with its two binding sites than IgM with its ten. A set of equations, similar to
Eqs. (6) and (7), are written down for the concentration of free
IgG and bound IgG [34]. However, Eq. (7) is replaced by three
equations to describe the three types of bound IgG's (see Fig. 4).
A diffusion-reaction equation for the free hapten should also be
included, but on physical grounds it has been argued that the
free hapten concentration, for the range of interesting concentrations, can be set equal to the total concentration. Solutions
have been considered in the two limits of single site attachment
and multisite attachment to the RBC. For single site attachment
the plaque radius can be shown to obey the following equation
[34]:

$$N = \frac{2KE}{1 + KH} \; \frac{S}{4\pi Dh} \; E_1 (R^2/4D^*t) \tag{13}$$

where $D^* = D/[1 + 2K\rho_0/(1 + KH)]$.

Note that in the limit that $KH \gg 1$, Eq. (13) becomes independent of K . For this range of H values, plaque inhibition

curves yield information about the distribution of emission rates and say nothing about the affinity. This will be true for IgM plaques as well.

A strong effort, primarily by DeLisi, has been made to define conditions under which plaque inhibition curves do yield information about the affinity distribution [34-37]. I will not discuss the conditions here except to note that a necessary condition is that the antibodies form long-lived multisite attachments at the RBC surface. Since immunologists almost never determine the hapten density they work at, it is difficult to know what limit is appropriate. To site an example, recently there has been a report, based on plaque inhibition studies, which claims to have observed a maturation in the IgM response at the cellular level [37]. This would be a most interesting result since there is much conflicting data on the subject. However, in none of the experiments on the maturation of IgM is the hapten density known. In light of Eq. (13) it may very well be that the conflicting results are due to differences in the hapten densities used, which causes some experimenters to unknowingly measure properties of the secretion rate rather than the affinity distribution.

Immunologists are beginning to realize that plaque inhibitions studies do not always reflect affinity distributions [38, 39]. However, as yet, there is no assay for affinity at the single cell level to take its place.

A new extension of the plaque technique has recently been suggested which involves developing plaques in the presence of an electric field [30]. Since antibodies are generally charged at pH = 7.4 (AFC must be maintained at physiological conditions), the plaques will become cigar shaped due to transport by the field. [The equations that describe plaque growth in an electric field are obtained by simply adding a transport term, $-v\partial c/\partial x$, to the right side of Eq. (6), where $v = \mu E$: μ is the antibody

mobility and E the magnitude of the external electric field.]
The theory predicts that from the plaque shape the mobility of
the antibodies produced by the AFC can be determined. This measurement is of interest since by determining the number of distinct
antibody mobilities a lower bound on the number of responding
clones may be obtained.

III.   THE MODELS

In this section I will concentrate on two mathematical models
which attempt to answer the question, "what happens to an animal
after it is injected with antigen?" First I will consider the
clonal selection model of Bell [9,40,41] and then the optimal
control model of Perelson, Mirmirani and Oster [42]. It is natural to discuss these two models together since, although their
approaches are quite different, they complement each other quite
nicely.

A.   A Model of Clonal Selection

In the theory of clonal selection a particular animal early
in its development acquires a large repertoire of B-cells. Different B-cells in the repertoire are committed to different antibodies, i.e., the progeny of a given B-cell secrete antibodies
whose binding sites are identical to the sites of the antibodies
that were imbedded in the surface of the parent B-cell. An antigen can trigger a number of different B-cells to proliferate by
binding to the surface antibodies (sometimes called receptor
molecules) of the B-cells. (The binding is necessary, but not
sufficient for triggering. See the previous paper.) The initial
B-cell, which is a small lymphocyte that does not secrete antibody, once triggered will proliferate into large lymphocytes
which secrete, and exhibit on their surface, the same antibody
as the original B-cell.[*] Large lymphocytes can continue to

----

[*] There is some evidence [43-47], although the interpretation is
highly controversial [48,49], that not all members of the same
clone produce antibodies with identical binding sites.

oliferate or they can differentiate (in general, the differen-
ation is not symmetric) into plasma cells and memory cells.
asma cells are terminal cells that are the most copious pro-
cers of antibodies. The memory cells are similar to the orig-
al small lymphocyte and in Bell's model, are taken to be iden-
cal with it.

Bell writes down differential equations for the circulating
tibody concentration, the total antigen concentration, and the
ur cell populations (small lymphocytes, called target cells in
ll's model, large lymphocytes, plasma cells and memory cells).
ese equations are coupled because the antigen binds to the
lls, the cells produce antibody, and the antibody binds to the
tigen. To calculate the binding, equilibrium is assumed and
e law of mass action used. The biological assumptions deter-
ne the form of the equations, while the predictions of the
del are obtained by numerically solving them.

Into the model must go the conditions for the triggering of
all lymphocytes, for the proliferation of large lymphocytes,
d for the differentiation of large lymphocytes into plasma and
mory cells. However, these conditions are not well understood
guesses as to the mathematical forms of the control functions
st be made. For the moment let us consider the cells that
ise from a single clone. The set of equations Bell writes down
r the four cell populations, the small lymphocytes (S), the
rge lymphocytes (L), the plasma cells (P) and the memory cells
), are*:

The equations Bell writes down are somewhat more general than
ese. His equations allow for the possibility of cell death
en a large fraction of the cells' receptors are bound to anti-
n. This possibility is neglected in the equations above. The
re general form used by Bell can account for a wider variety of
gh zone tolerance effects. These effects are briefly discussed
ter in this section.

$$\frac{dS}{dt} = -F \frac{S}{t'_S} + source \tag{14}$$

$$\frac{dL}{dt} = F \frac{S}{t'_S} + H \frac{L}{t'_L} - \frac{L}{T_L} \tag{15}$$

$$\frac{dP}{dt} = F_P(1 - H) \frac{L}{t'_L} - \frac{P}{T_P} \tag{16}$$

$$\frac{dM}{dt} = (1 - F_P)(1 - H) \frac{L}{t'_L} - \frac{M}{T_M} \tag{17}$$

where T_L , T_P and T_M are the mean lifetimes of large lymphocytes, plasma cells and memory cells respectively. t'_S is the
mean time for a stimulated small lymphocyte to become a large
lymphocyte and t'_L is the mean time for a large proliferating
lymphocyte to divide. t'_S , t'_L , F and H are all functions
of the antigen binding, the details of which will be discussed
shortly.[*]

Usually the memory cells are taken to be identical with the
small lymphocytes. The source of small lymphocytes in Eq. (14)
is set equal to $(1 - F_P)(1 - H)L/t'_L$ and $1/T_M$ is set equal to
zero.

One of the basic assumptions of the model is that it is the
antigen which controls triggering, proliferation and differentiation. Since the antigen interacts with the small and large
lymphocytes by binding to the surface receptors, Bell takes t'_S ,
t'_L , F and H to be functions of r', the fraction of occupied
receptor sites. He also takes them to be functions of two parameters f_{min} and f_{max} which are constants. The parameters are
introduced in such a way that if $r' \gg f_{max}$ or $r' \ll f_{min}$,
essentially no triggering, proliferation or differentiation can
occur. If patching is a necessary step for these events, this

---

[*] In this section H is a particular function introduced by Bell
and defined by Eq (22). It no longer represents a hapten concentration.

ype of behavior must be built into the model since at very large
d very small antigen concentrations, patching cannot occur (see
e previous paper).

Although the number of large lymphocytes that differentiate
to either plasma or memory cells is determined by the antigen
ncentration, in Bell's model the fraction that becomes plasma
lls, F_P , and the fraction that becomes memory cells, $(1-F_P)$,
e independent of the antigen concentration. For a particular
tigen F_P is a constant. (In a recent model of Bruni et al.,
ich is similar to Bell's model, F_P is taken equal to r' [50].)

The particular forms Bell has chosen for t_S' and t_L' are
follows:

$$t_S' = \frac{T_S'}{1 - G} \tag{18}$$

$$t_L' = \frac{T_L'}{1 - G} \tag{19}$$

ere

$$G = \frac{r'\delta}{1 - r' + r'\delta} \tag{20}$$

d $\delta = (1 - f_{max})/f_{max}$. T_S' and T_L' are constants equal to
e minimum time for a small lymphocyte to become a large lympho-
te and a large lymphocyte to divide.

When the antigen concentration is very large so that
$\gg f_{max}$; $r'\delta \gg 1$, and $G \simeq 1$. In this limit, the times
r stimulation and proliferation become infinite and there is no
mune response. The nonresponse of the immune system to a high
se of antigen is an observed property of the immune system
own as high zone tolerance. (For an introduction to tolerance
fects see p. 492-502 of ref. [6]. To see how Bell's complete
del treats tolerance see ref. [9].)

For F and H Bell has chosen the following forms:

$$F = \frac{r'}{r' + f_{min}} \tag{21}$$

$$H = \frac{r' - f_{min}}{r' + f_{min}} \quad . \tag{22}$$

When the antigen concentration is very low so that
$r' \ll f_{min}$, $F \simeq 0$. In this limit the antigen concentration is
too low to cause an immune response, as can be seen from Eq. (14)
and small lymphocytes are not stimulated.

Consider a set of small lymphocytes, each characterized by a
different binding affinity. If the antigen concentration decreases with time, those with low affinities will have $r' \ll f_{min}$
while those with high affinities will have $r' \gtrsim f_{min}$. As time
increases and the antigen concentration drops the model predicts
more and more of the low affinity cells will go unstimulated.
This will lead to an increase in the average affinity with time,
i.e., a maturation of the immune response. Similar arguments
concerning f_{max} and the function (1 - G) show that as time
increases stimulation of small lymphocytes with higher affinities
occur. Bell's model gives good agreement with experimentally observed maturation [41].

The function H determines whether a large lymphocyte will
proliferate or differentiate. When $r' \gg f_{min}$, $H \simeq 1$ and
$F \simeq 1$. In this limit large lymphocytes proliferate but do not
differentiate. It is only when the antigen concentration drops
to values where H no longer is essentially equal to one that
differentiation into plasma and memory cells becomes significant.
When I discuss the optimal control model of Perelson et al, you
will see that with this choice of H , the immune response is
very close to being optimal.[*]

---
[*] A precise definition of "optimal" is presented when the optimal
control model is discussed.

When Eqs. (14-17), along with the equations for the total antibody and antigen concentrations, and the chemical equilibrium condition (these last two equations and the chemical equilibrium condition will not be discussed in this paper, see [40] for a complete discussion) are solved numerically for a single initial antigen dose and reasonable parameter values, the following is predicted: large lymphocytes proliferate with very little differentiation for a long initial period, then switch to differentiation with very little proliferation within a short time period (see Fig. 2 in [40]). Bell has considered a variety of other cases as well.

Up to now I have only discussed B-cells (the small and large lymphocytes and the plasma and memory cells in Bell's model are all B-cells). However, in order for an animal to produce a humoral response against most antigens, it must possess at least two types of viable lymphocytes, B-cells and T-helper cells. However, for a certain class of antigens, called T-independent antigens, a humoral response will take place in the absence of T-helper cells. These antigens all have very similar structure, being large polymers with many copies of the same antigenic determinant. For T-independent antigens Bell's model seems quite complete. For such antigens F_p should probably be set equal to one since there is some experimental evidence that memory cells are not produced [51]. For T-dependent antigens Bell's model is not complete (indeed the model was formulated before the discovery of the helper function that T-cells provide; Bruni's model also does not treat T-cells [50]). The T-cells play an important role in controlling the immune response, but exactly how to incorporate them into Bell's model is not clear. The details of how T- and B-cells interact in the presence of antigens is a major area of immunological study and possibly in the near future sufficient knowledge will be gained to model interacting T- and B-cell populations.

B.   An Optimal Control Model

Perelson, Mirmirani and Oster have posed a most interesting question: "what is the optimal strategy available to the immune system for responding to an antigenic challenge?" [42]. For the case of T-independent antigens they have answered this question.

Although there is no a priori reason which assures that the immune system responds to antigen in an optimal way, the mammalian immune system has been evolutionarily static for a long time It is quite possible that the system has become static because it has evolved to the point where it performs optimally. I shall assume this is true for the remainder of the discussion.

In the model an animal is injected with a T-independent anti gen which causes a number of small lymphocytes to be transformed into L_0 large lymphocytes. The dynamics of the small lymphocyte population is not considered. The large lymphocytes (L) can proliferate, or differentiate into plasma cells (P). Memory cells do not arise because the antigen is T independent [51]. The large lymphocytes secrete antibodies (A) at a rate S while plasma cells secrete at a rate γS, where $\gamma > 1$ and S is measured in antibodies/sec.

The simplest set of equations treated in the model are the following:

$$\frac{dA}{dt} = S(L + \gamma P) \tag{23}$$

$$\frac{dL}{dt} = u(t) \frac{L}{t_L'} - [1 - u(t)] \frac{L}{t_d'} - \frac{L}{T_L} \tag{24}$$

$$\frac{dP}{dt} = [1 - u(t)] \frac{L}{t_d'} - \frac{P}{T_P} \tag{25}$$

where $0 \leqslant u(t) \leqslant 1$. $u(t)$ is the fraction of cells that remain large lymphocytes and $1 - u(t)$ is the fraction that becomes plasma cells. T_L and T_P are the mean lifetimes of large

ymphocytes and plasma cells respectively. t_L' is the mean time
or proliferation of large lymphocytes and t_d' is the mean time
or differentiation of large lymphocytes into plasma cells.

Before discussing the optimal control problem, it is useful
o compare these equations with the equations of Bell's model.
L and T_P are the same in both models. In Bell's model
$_L' = t_d'$. If I set $t_L' = t_d'$ in Eq. (24), then

$$\frac{dL}{dt} = [2u(t) - 1] \frac{L}{t_L'} - \frac{L}{T_L} \quad . \qquad (26)$$

In Bell's model the proliferation time depends on the anti-
en concentration, i.e., $t_L' = T_L'/(1 - G)$, and therefore, it is a
unction of time; in the optimal control model the proliferation
ime is a constant. By comparing Eq. (26) with Eq. (15) it fol-
ows that: $H = 2u - 1$.

The source of large lymphocytes are the triggered small lym-
hocytes. In Bell's model an explicit expression for this source
s given [the first term on the right side of Eq. (15)] and small
ymphocytes seed the large lymphocytes in decreasing amounts
hrough at least the early course of the response. In the optimal
ontrol model the source of large lymphocytes enters only at the
eginning of the response as an initial condition; at time $t = 0$,
$= L_0$, $P = 0$ and $A = 0$. Finally, unlike Bell's model, in
he optimal control model the interaction of the antigen with the
ntibody and lymphocytes is ignored. The first model of Perelson
nd his colleagues is a considerably simplified description of
he immune response. The advantage of such simplification is that
t leads to a most tractable optimal control problem.

The optimal control problem has been set up in the following
ay. The immune response is said to be optimal if a fixed amount
f antibody A^* is secreted in a minimum time T . This defini-
ion of optimal arises from the assumption that the antigen will
e eliminated from the animal if enough antibody is secreted in a

short enough time. The problem is to find the $u(t)$ that will
minimize the time required to secrete A^*. Formally the optimization criterion is

$$\min_{u} \int_0^T dT \qquad (27)$$

$$A(T) = A^* \quad .$$

This optimization criterion along with Eqs. (23)-(25) make
up a bilinear optimal control problem. Perelson, Mirmirani and
Oster have used Pontryagin's maximum principle to compute the
optimal control strategy, $u^*(t)$. They find the $u^*(t)$ can
take on only the two values 0 or 1 , which is known as "bang-
bang" control. This means that for this problem no graded response in the control is more efficient in the sense of the optimization criterion than an all or none type switch. Whether or
not a switch occurs depends only on the following parameters:
γ , t_L' , and t_d' and A^*. The time at which the switch occurs
and the total time needed to produce A^* depend on the full
parameter set $(\gamma, T_L, t_L', t_d', A^*, D)$.

If $1/t_L' \geqslant (\gamma - 1)/t_d'$, $u^*(t) = 1$ for $0 \leqslant t \leqslant T$; it is
optimal to proliferate without ever differentiating. It is easy
to understand why this is so by considering the case when $t_L' = t_d'$
Then this solution holds if $2 > \gamma$. Since on division a large
lymphocyte divides into two cells but on differentiation it is
transformed into only a single plasma cell, unless the plasma
emission rate is more than twice that of the large lymphocytes,
i.e., $\gamma S > 2S$, it never pays for a large lymphocyte to differentiate.

If $1/t_L' < (\gamma-1)/t_d'$ there are two possible solutions. If
A^* is small, $u^*(t) = 0$ for $0 \leqslant t \leqslant T$; it is optimal for the
L_0 large lymphocytes to immediately differentiate into plasma

ells. The value of A* below which this solution holds requires
olving a messy transcendental equation. However it is easy to
how that if $A* > SL_0(1 + \gamma T_P/t'_d)(1/t'_d + 1/T_L)$ the solution can
ever hold. The right side of the inequality is just the maximum
mount of antibody the large lymphocyte and plasma cell popula-
ions can produce if at time $t = 0$ the switch occurs. For this
ase Eqs. (24) and (25) become

$$\frac{dL}{dt} = - \left(\frac{1}{t'_d} + \frac{1}{T_L} \right) L \qquad (28)$$

$$\frac{dP}{dt} = \frac{L}{t'_d} - \frac{P}{T_P} \quad . \qquad (29)$$

hese equations can easily be solved for $P(t)$ and $L(t)$. The
otal antibody produced is then just

$$A(t) = S \int_0^t P(t')dt' + S \int_0^t L(t')dt'. \qquad (30)$$

ote that although the control is "bang-bang" and the switch
ccurs at time $t = 0$, the large lymphocyte population does not
nstantly go over into plasma cells, but decays exponentially
Eq. (28)].

   If the upper limit in Eq. (30) is set equal to infinity the
esult is

$$A(\infty) = SL_0(1 + \gamma T_P/t'_d)/(1/t'_d + 1/T_L) \qquad (31)$$

rom which the inequality follows.

   For reasonable values of the parameters it can be shown that
* $\gg A(\infty)$ [42]. When this inequality holds, $u*(t) = 0$ for
) $\leq t < t*$, and $u*(t) = 1$ for $t* \leq t \leq T$, where $t*$ is the
witching time. Thus, in most immune responses to T-independent
ntigens I expect that large lymphocytes will proliferate until

time t* with no differentiation into plasma cells, and then
after time t* large lymphocytes will differentiate with no proliferation. In Bell's model the behavior of these two cell populations is very close to optimal. (To be optimal (H - 1)/2
would have to be a step function.)

I have discussed only the simplest case that Perelson,
Mirmirani and Oster have considered. They have extended their
work to more realistic models [42]. Still to be formulated are
models which elucidate the controls for stimulation of small
lymphocytes and the production of memory cells. The latter model
requires the introduction of T-cell populations. I am sure a
number of other interesting immunological quesitons can be formulated in terms of optimal control problems.

In Bell's model guesses had to be made as to the control
functions which regulate stimulation, proliferation and differentiation. In a complete optimal control model these are solved
for. It is the optimization criteria which are speculative. The
interaction of models such as these seems to be a promising way
to improve our understanding of the immune response.

Acknowledgements

I thank Alan Perelson and George Bell for their comments
and criticisms.

REFERENCES

1. W. O. Weigle, "Cyclical Production of Antibody as a Regula-
    tory Mechanism in the Immune Response," Adv. Immun. 21 (1975)
    87-111.

2. Y. O. Kim and F. Karush, "Equine Anti-Hapten Antibody - VIII.
    Isoelectric Fractions of IgM and 7S Anti-Lactose Antibody,"
    Immunochem. 11 (1974) 147-152.

3. H. W. Kreth and A. R. Williamson, "The Extent of Diversity
    of Anti-Hapten Antibodies in Inbred Mice: Anti-NIP (4-
    hydroxy-5-iodo-nitro-phenacetyl) Antibodies in CBA/H Mice,"
    Eur. J. Immunol. 3 (1973) 141-147.

. Y. T. Kim, T. P. Weblin and G. W. Siskind, "Distribution of
   Antibody Affinities - II. Fractionation of Antibody with
   Respect to its Hapten Binding Affinity," Immunochem. 11
   (1974) 685-690.

. B. Goldstein, "Theory of Hapten Binding to IgM: The Question
   of Repulsive Interactions Between Binding Sites," Biophy.
   Chem. 3 (1975) 363-367.

. B. D. Davis, R. Dulbecco, H. Eisen, H. Ginsberg and W. B.
   Wood, Microbiology, 2nd Edition, Hoeber Med. Div. - Harper
   & Row, New York (1974).

. R. S. Farr, "A Quantitative Immunochemical Measure of the
   Primary Interaction Between I*BSA and Antibody," J. Infect.
   Dis. 103 (1958) 239-262.

. Y. T. Kim, S. Kalver and G. W. Siskind, "A Comparison of the
   Farr Technique with Equilibrium Dialysis for Measurement of
   Antibody Concentration and Affinity," J. Immunol. Methods 6
   (1975) 347-354.

. G. I. Bell, "Mathematical Model of Clonal Selection and Antibody Production. III. The Cellular Basis of Immunological
   Paralysis," J. Theor. Biol. 33 (1971) 379-398.

. H. N. Eisen and G. W. Siskind, "Variations in Affinities of
   Antibodies During the Immune Response," Biochem. 3 (1964)
   996-1008.

. G. W. Siskind and B. Benacerraf, "Cell Selection by Antigen
   in the Immune Response," Adv. Immun. 10 (1969) 1-50.

. G. W. Siskind, "Heterogeneity of Antibody-Binding Affinity,"
   in Mammalian Cells: Probes and Problems, Ed. C. R. Richmond,
   D. F. Peterson, P. F. Mullaney and E. C. Anderson (1975)
   277-283.

. G. I. Bell, "Model for the Binding of Multivalent Antigen
   to Cell," Nature 248 (1974) 430-431.

. C. DeLisi and A. Perelson, "The Kinetics of Aggregation
   Phenomena. I. Minimal Models for Patch Formation on Lymphocyte Membranes," J. Theor. Biol. (in press).

. T. P. Weblin and G. W. Siskind, "Distribution of Antibody
   Affinities: Technique of Measurement," Immunochem. 9 (1972)
   987-1011.

16.   T. P. Weblin, Y. T. Kim, F. Quagliata and G. W. Siskind,
        "Studies on the Control of Antibody Synthesis. III. Change
        in Heterogeneity of Antibody Affinity During the Course of
        the Immune Response," Immunology 24 (1973) 477-491.

17.   Y. T. Kim and G. W. Siskind, "Studies on the Control of Anti-
        body Synthesis. VI. Effect of Antigen Dose and Time After
        Immunization on Antibody Affinity and Heterogeneity in the
        Mouse," Clin. Exp. Immunol. 17 (1974) 329-338.

18.   T. P. Weblin and G. W. Siskin, "Effect of Tolerance and
        Immunity on Antibody Affinity," Transplant. Rev. 8 (1974)
        104-136.

19.   C. Bruni, A. Germani, G. Koch and R. Strom, "Derivation of
        Antibody Distribution of Antibody Affinities in the Immune
        Response," J. Theor. Biol. (in press).

20.   N. K. Jerne, A. A. Nordin and C. Henry, "The Agar Plaque
        Technique for Recognizing Antibody Producing Cells," in
        Cell Bound Antibodies, Wistar Institute Press, Philadelphi
        (1963) 109.

21.   N. K. Jerne, C. Henry, A. A. Nordin, H. Fuki, A.M.C. Koros
        and I. Lefkovits, "Plaque Forming Cells: Methodology and
        Theory," Transplant. Rev. 18 (1974) 130-191.

22.   A. J. Cunningham and A. Szenberg, "Further Improvements in
        the Plaque Technique for Detecting Single Antibody-Forming
        Cells," Immunology 14 (1968) 599-600.

23.   J. H. Humphrey, "Haemolytic Efficiency of Rabbit IgG Anti-
        Forssman Antibody and its Augmentation by Anti-Rabbit IgG,
        Nature 216 (1967) 1295-1296.

24.   J. S. Ingraham and A. J. Bussard, "Application of Localize
        Hemolysin Reaction to Specific Detection of Antibody Formi
        Cells," J. Exp. Med. 119 (1964) 667-678.

25.   G.J.V. Nossal, A. E. Bussard, H. Lewis and J. C. Mazie, "I
        Vitro Stimulation of Antibody Formation by Peritoneal Cell
        J. Exp. Med. 131 (1970) 894-916.

26.   G.J.V. Nossal and H. Lewis, "Functional Symmetry Amongst
        Daughter Cells Arising in Vitro From Single Antibody Formi
        Cells," Immunology 20 (1971) 739-753.

27.   C. P. DeLisi and G. I. Bell, "The Kinetics of Hemolytic
        Plaque Formation," Proc. Nat. Acad. Sci. USA 71 (1974) 16-

28. B. Goldstein, C. DeLisi and J. Abate, "Immunodiffusion in
     Gels Containing Erythrocyte Antigen. I. Theory for Diffusion
     from a Circular Well," J. Theor. Biol. 52 (1975) 317-334.

29. C. DeLisi, "The Kinetics of Hemolytic Plaque Foramtion - V.
     The Influence of Geometry on Plaque Growth," J. Math. Biol.
     2 (1975) 317-331.

30. B. Goldstein and A. Perelson, "The Electrophoretic Hemolytic
     Plaque Assay - Theory," Biophys. Chem. (in press).

31. C. L. Hornick and F. Karush, "Antibody Affinity - III. The
     Role of Multivalence," Immunochem. 1 (1972) 325-340.

32. B. Anderson, "Studies on the Regulation of Avidity at the
     Level of the Single Antibody-Forming Cell," J. Exp. Med.
     132 (1970) 77-88.

33. G. W. Miller and D. Segre, "Determination of Relative Affin-
     ity and Heterogeneity of Mouse Anti-DNP Antibodies by a
     Plaque-Inhibition Technique," J. of Immunol. 109 (1972) 74-83.

34. C. DeLisi and B. Goldstein, "The Kinetics of Hemolytic Plaque
     Formation - II. Inhibition of Plaques by Hapten," J. Theor.
     Biol. 51 (1975) 313-335.

35. C. DeLisi, "The Kinetics of Hemolytic Plaque Formation -
     III. Inhibition of Plaques by Antigen," J. Theor. Biol. 51
     (1975) 337-345.

36. C. DeLisi, "The Kinetics of Hemolytic Plaque Foramtion - IV.
     IgM Plaque Inhibition," J. Theor. Biol. 52 (1975) 419-440.

37. E. A. Goidl, J. J. Barondess and G. W. Siskind, "Studies in
     the Control of Antibody Synthesis - VII. Changes in Affinity
     of Direct and Indirect Plaque-Forming Cells with Time After
     Immunization in the Mouse: Loss of High Affinity Plaques
     Late After Immunization," Immunology 29 (1975) 629-641.

38. J. R. North and B. A. Askonas, "Analysis of Affinity of
     Monoclonal Antibody Responses by Inhibition of Plaque-Form-
     ing Cells," Eur. J. Immunol. 4 (1974) 361-366.

39. E. A. Goidl, G. Birbaum and G. W. Siskind, "Determination of
     Antibody Avidity at the Cellular Level by the Plaque Inhibi-
     tion Technique: Effect of Valence of Inhibitor," J. of
     Immunol. Methods 8 (1975) 47-52.

40. G. I. Bell, "Mathematical Model of Clonal Selection and
     Antibody Production," J. Theor. Biol. 29 (1970) 191-232.

41. G. I. Bell, "Mathematical Model of Clonal Selection and
     Antibody Production. II.," J. Theor. Biol. 33 (1971) 339-378

42. A. S. Perelson, M. Mirmirani and G. F. Oster, "Optimal Strat
     egies in Immunology," to be published.

43. A. J. Cunningham and S. A. Fordham, "Antibody Cell Daughters
     can Produce Antibody of Different Specificities," Nature 250
     (1974) 669-671.

44. A. J. Cunningham and L. M. Pilarski, "Generation of Antibody
     Diversity - I. Kinetics of Produciton of Different Antibody
     Specificities During the Course of an Immune Response," Eur.
     J. Immunol. 4 (1974) 319-326.

45. A. J. Cunningham and L. M. Pilarski, "The Generation of
     Antibody Diversity - II. Plaque Morphology as a Simple Marke
     for Antibody Specificity at the Single-Cell Level," Eur. J.
     Immunol. 4 (1974) 757-761.

46. L. M. Pilarski and A. J. Cunningham, "The Generation of
     Antibody Diversity - III. Variation in the Specificity of
     Antibody Produced Within Single Clones of Antibody-Forming
     Cells in Vitro," Eur. J. Immunol. 4 (1974) 762-767.

47. L. M. Pilarski and A. J. Cunningham, "Generation of Antibody
     Diversity - IV. Variation Within Single Clones of Antibody-
     Forming Cells Developing in Vivo," Eur. J. Immunol. 5 (1975)
     10-16.

48. B. Goldstein, "Effect of Antibody Emission Rates on Plaque
     Morphology," Nature 253 (1975) 637-639.

49. C. DeLisi and G. I. Bell, "Plaque Morphology as an Antibody
     Specificity Marker: An Analysis of the Physicsl Chemical
     Foundations of the Method," Immunochem. 13 (1976) 21-28.

50. C. Bruni, M. A. Giovenco, G. Koch and R. Strom, "A Dynamical
     Model of Humoral Immune Response," Math. Bioscience 27 (1975)
     191-211.

51. W. E. Paul, M. Kaapf and D. E. Mosier, "Activation and Tol-
     erance Induction in DNP-specific B-cells: Analysis with Three
     Distinct DNP-carrier Conjugates," in Immunological Tolerance
     Mechanisms and Potential Therapeutic Applications, ed. D. H.
     Katz and B. Benacerraf, Academic Press, New York (1974) 141-
     161.

THEORETICAL DIVISION   DEPARTMENT OF PHYSICS
UNIVERSITY OF CALIFORNIA   FAIRLEIGH DICKINSON UNIVERSITY
LOS ALAMOS SCIENTIFIC LABORATORY   TEANECK, NEW JERSEY 07666
LOS ALAMOS, NEW MEXICO 87545

CURRENT PROBLEMS IN PATTERN FORMATION

H. G. Othmer

INTRODUCTION

One of the oldest and most intriguing problems in
neoretical biology concerns the origin of spatial pattern
a developing systems. In Wolpert's formulation [1], the
roblem is that of 'assigning specific states to an ensemble
: cells, whose initial states are relatively similar, such
nat the resulting ensemble of states forms a well-defined
patial pattern'. The first significant mathematical analysis
? this problem was done by Turing [2], who originated what
s currently called the reaction-diffusion theory of pattern
ormation. Turing's theory is built around the remarkable
act, which he first proved, that a spatially uniform
cationary state of a reacting mixture can be unstable to
patially nonuniform disturbances if reaction and diffusion
nteract appropriately. The theory then envisions that as
ertain slowly-varying kinetic or transport coefficients
ross critical (bifurcation) values, the uniform state loses
cability and a spatially nonuniform state emerges.
epending on the nature of the instability, the resulting
pnuniform state may be steady or time dependent. If one or
pre of the chemicals in this spatially nonuniform concentra-
ion pattern activates transcription of a gene that codes
pr a key enzyme or structural protein, a nonuniform pattern
f cell differentiation will result. Such an interpretation

Supported in part by NIH Grant # GM 21558

has been used to explain the origin of insect bristle
patterns [3].

In the past decade there has been a renewal of interest
in the theoretical aspects of Turing's theory, stimulated in
part by Gmitro and Scriven's work [4], and in part by the
wide variety of spatio-temporal patterns observed in the
Belousov-Zhabotinskii reaction [5]. Linear stability analys
the first step in any analysis of pattern formation, has
been worked out in detail for two- and three-component
systems with arbitrary kinetic mechanisms and diffusion
matrices [6]. Such analysis is useful in studying the onset
of instability and for gaining insight into how the differen
types of instability are produced by the interaction of
reaction and diffusion. However, to predict what spatial
pattern ultimately evolves from an instability requires a
nonlinear analysis, and here one cannot expect results of th
scope available in the linear theory [7]. Certainly more
analysis of specific kinetic mechanisms, such as that done
in [8],is needed before any general conclusions on the
evolution of systems near their bifurcation points emerge.
When parameters are far from their bifurcation values, there
is usually no alternative to numerical solution of the
governing equations. An indication of the complexity of
patterns possible in reaction-diffusion systems is given by
the computational results of Gierer and Meinhardt [9].

The outlines of a comprehensive nonlinear theory of
pattern formation from steady uniform states are emerging
from analyses such as the foregoing and those done in the
context of ecological problems [10]. Some important
problems that remain are the following.

(1)   Linear stability analysis answers the question
       of how reaction, diffusion and system geometry
       or topology interact to produce instability.
       A complete nonlinear theory should predict, for
       example, how these factors govern the direction
       and stability of the bifurcating solutions.
       One general result will be given in a later
       section.

(2) Closely related to the preceding point are the
     problems of bifurcation from multiple eigenvalues
     and of secondary bifurcation when there is a pair
     of nearly-degenerate eigenvalues [11]. These in
     turn are related to the problem of pattern
     selection when more than one type of pattern is
     predicted by linear analysis [12]. Results for
     these problems will illustrate how a succession
     of instabilities can be used to generate
     increasingly complex spatial patterns.

(3) In Turing's work and virtually all subsequent work,
     the only mode of transport considered is diffusion.
     Furthermore, the model systems all deal with
     structureless, tightly coupled cells or their
     continuum analogs. Some future work should be
     directed toward (a) the analysis of other modes
     of transport, (b) more realistic models of cell
     and tissue structure [13], and (c) networks of
     cells that communicate only indirectly via the
     external medium.

While significant progress has been made toward
predicting the emergence of spatial pattern from uniform
steady states, the theory is still in its infancy when the
underlying chemical dynamics are time-periodic. Even linear
analysis of the interaction between reaction and transport
is difficult in such cases, because the equations are
non-autonomous and the periodic solution is rarely known in
analytical form. Nonetheless, the following examples suggest
that the problem is of sufficient importance to warrant
detailed investigation.

It is widely recognized that biological systems are
periodic at virtually every level of organization from the
sub-cellular to the organismic level. The most carefully
studied biochemical oscillations are the glycolytic
oscillations in yeast cells [14], observable in single cells,
in intact cell suspensions where cells communicate
directly, and in cell-free extract. There is as yet no

direct evidence that links glycolytic oscillations with any
morphogenetically significant process, but many have speculat
that similar oscillations may be important in circadian
rhythms [15]. Populations of interacting oscillators have
been studied in this context [16].

The second class of examples in which the dynamics are
time-periodic comprises multinucleate cells obtained by
fusion [17], the true slime molds such as Physarum
polycephalum [18], and the colonial fungi [19,20]. In
multinucleate cells and in Physarum, the underlying periodic
is that of the mitotic cycle, while fungi undergo a more
complicated life cycle that involves differentiation. It
has been observed that in Physarum the contents of fused
plasmodia are mixed by a combination of diffusion and
cytoplasmic streaming and that nuclei in a fused plasmodium
divide synchronously provided the diameter is less than
about 15 cm [21]. In larger plasmodia, propagating waves of
mitosis can occur. As the paper by Kauffman in this volume
shows, this system is an ideal one for testing hypotheses
about the basis for the periodicity in the mitotic cycle.

More complicated spatial patterns are often observed in
various fungi. These include concentric rings and spirals
[20] very similar to those observed in the Belousov-
Zhabotinskii reaction. In the ascomycete Chaetomium robustu
the alternating light and dark zones seen in a growing color
are transparent zones of scantily branched hyphae alternatin
with dense zones of intense ramification. In relatively
small colonies these dense zones are very uniform around the
circumference, which indicates a high degree of synchrony
between adjacent hyphae, at least in the latter stage of the
life cycle. The mode of communication between hyphae is not
yet known, but a plausible hypothesis is that synchronizatio
is maintained by transport of some 'messenger' molecule.
The conjecture that transport is by diffusion is supported
by the fact that the dense zones become increasingly
irregular as the colony expands [20]. Moreover, it has not
been established that the life cycle is controlled by a
biochemical oscillator but it could well be. Granting this,

ie observed patterns are amenable to analysis using a
action-transport model. A highly-simplified model, which
netheless gives rise to spatial patterns similar to those
served, is due to Pavlidis [22].

In addition to the experimental examples, there are
imerous theoretical models in which interaction between cells
.th periodic dynamics plays a role. Waddington [23]
iggested that entrainment of non-oscillatory cells by
icillatory cells could constitute a mechanism of tissue
iduction in embryology. Goodwin and Cohen [24] have
instructed a model of development in which oscillatory cells
i a developing tissue are entrained to the frequency of
calized pacemakers by periodically-propagating waves.
ie local phase difference between a fast wave and a slow
.ve provides positional information and thereby governs
cal differentiation.

Burton and Canham [25] have recently proposed a model
r contact inhibition of cell division based on biochemical
cillators coupled by diffusion. This model stems from the
servations by Lowenstein and Kanno [26] that intercellular
mmunication in certain tumor cells is very slight compared
th communication in their normal counterparts. The main
potheses of the Burton-Canham model are that there is a
y substance involved in contact inhibition that diffuses
tween cells and whose concentration within each cell
cillates harmonically in time, with a period that varies
om cell to cell. They propose that contact inhibition
sults when cells communicate freely and the level of the
y substance remains below a threshold for initiation of
e mitotic cycle. Their major result is that the amplitude
 the oscillation can be suppressed by virtue of coupling
tween neighboring cells. Despite its apparent success,
veral criticisms of the model can be raised. Firstly,
llular communication has no affect on the dynamics of the
cillator but merely provides for leakage of the key
bstance between neighboring cells. This forces the authors
 assume that all cells have different frequencies so as to
eclude synchronization. Secondly, it is generally true

that only small-amplitude oscillations are satisfactorily
approximated by sinusoids and as a result, the control
system proposed is very sensitive to small concentration
changes. Both of these criticisms can be vitiated by
postulating that the transported species is directly involve
in the dynamics of the underlying oscillator, as Kauffman
and Wille [27] assume in their model of mitosis. Nonetheles
the fundamental idea of Burton and Canham is attractive and
warrants further investigation.

The common features of most of the foregoing
experimental examples and theoretical models are (i) the
dynamics of individual 'units' are periodic or capable of
being entrained by a periodic signal and (ii) the units are
coupled, either directly via diffusion or active transport
across cell membranes, tight junctions or within a
plasmodium, or indirectly, by transport of a chemical specie
through an external medium in contact with all cells. In
all the examples one can ask (i) under what conditions do th
coupled units synchronize into a single collective mode and
(ii) what properties of the internal (biochemical) dynamics
and of the coupling produce nonuniform spatio-temporal
patterns? These questions are addressed in the following
sections.

The outline for the remainder of the paper is as
follows. Section II describes the general hypotheses on the
chemical kinetics. Because the analysis and exposition is
simplest when there are only two chemical species, we
restrict attention to this case. Most of the results are
extendable to systems involving more species at the expense
of added algebraic detail. In Section III we use a
continuum model of coupled cells to derive conditions on
kinetic and transport parameters that provide the answer to
the first question of the preceding paragraph. The conditic
are stringent and will generally be met only in small
systems. Weaker conditions will suffice to guarantee that
a system in a spatially-uniform periodic state remains there
in the face of small random concentration disturbances.
These are derived in Section IV. A partial answer to the

econd question posed is given in Section V and several
ossible topics for further investigation are suggested in
he concluding section.

I.   THE CHEMICAL KINETICS

No specific kinetic mechanism involving the two active
pecies will be postulated here; instead we simply assume
hat the kinetics are described by a smooth nonlinear
unction $R(c,p)$ and that the equation of change for a
niform system is

$$\frac{dc}{dt} = R(c,p). \tag{1}$$

ere

$$c = \begin{pmatrix} c_1 \\ c_2 \end{pmatrix} \qquad R(c,p) = \begin{pmatrix} R_1(c_1,c_2,p) \\ R_2(c_1,c_2,p) \end{pmatrix}$$

nd p is a positive scalar parameter. This parameter
ight, for instance, be the concentration of a slowly-varying
ubstrate. Throughout we assume that R is such that the
olution of (1) for positive initial values remains non-
egative and bounded for all $t>0$. In addition to these
tandard assumptions, we make the following hypotheses about
..

(H1)   For every $p>0$ there is a unique solution
c^* of $R(c,p)=0$.

(H2)   Let K be the matrix of the linearization of
(1) around c^*, <u>viz.</u>

$$\frac{d\zeta}{dt} = K\zeta$$

$$\zeta \equiv c-c^* \qquad K_{ij} \equiv \left(\frac{\partial R_i}{\partial c_j}\right) c_j = c_j^*$$

There are two parameter values p_0 and p_1, $0<p_0<p_1$,
such that both eigenvalues of K have non-negative
real parts for $p_0 \le p \le p_1$. The eigenvalues are complex
conjugates, $\alpha \pm i\beta$, in some neighborhood of p_0
and in some neighborhood of p_1. At p_0

$$\alpha(p_0) = 0, \qquad \frac{d\alpha}{dp}(p_0) > 0$$

while at p_1

$$\alpha(p_1) = 0, \qquad \frac{d\alpha}{dp}(p_1) < 0,$$

(H3) For any $p \varepsilon (p_0, p_1)$ there exists a unique
      periodic solution that is globally asymptotically
      stable, but for the steady state c^*. There
      are no periodic solutions for any other
      value of p.

It follows from (H2), (H3) and the Hopf theorem [28] that a
stable periodic solution emerges at p_0 and disappears at
p_1 as p increases through these values. It can happen
that $p_1 = \infty$.

The matrix K will have a pair of complex conjugate
eigenvalues with non-negative real part if and only if

$$\mathrm{tr}\, K = k_{11} + k_{22} \geq 0$$

and

$$(\mathrm{tr}\, K)^2 - 4 \det K = (k_{11} - k_{22})^2 + 4\, k_{12}k_{21} < 0.$$

The first condition requires that at least one of the species
be self-activating since at least one k_{ii} must be positive.
We shall specify that species 1 is self-inhibiting for all
p and so k_{11} is always negative. Therefore species 2
must be self-activating near p_0 and p_1 and in fact, it
follows from (H2) that k_{22} must be positive for all p in
an interval $[\hat{p}_0, \hat{p}_1]$ that contains $[p_0, p_1]$. We can allow
$\hat{p}_0 = 0$ and $\hat{p}_1 = \infty$. To satisfy the second condition it is
necessary that $k_{12}k_{21} < 0$. This means that the mutual
interaction of 1 and 2 near p_0 and p_1 must either be
that 1 activates 2 and 2 inhibits 1 or, that 1 inhibits 2 and
2 activates 1. One model reaction mechanism that fulfills
all the above hypotheses is the mechanism proposed by
Zhabotinskii, et. al. [29] for the Belousov-Zhabotinskii
reaction. This model is analyzed in [30]. A somewhat
simpler scheme that also has these properties arises from
models for glycolytic oscillations [31].

III.   GLOBAL STABILITY OF UNIFORM SOLUTIONS

Now suppose that we have N identical cells, in each
of which the kinetics for the two species of interest are
as just described. Further, suppose that the cells are
submerged in a bath and that both species can be exchanged
between cells. This exchange can occur in one of two ways.
In the first, communication between cells is indirect in that
each cell exchanges material only with the extracellular
milieu. This mode is used in yeast cell suspensions and
presumably in many other similar situations. We shall not
pursue an analysis of indirect communication here; suffice
it to say that when individual cells are oscillatory,
$(p\varepsilon[p_0,p_1])$, the oscillations can be suppressed *in vivo*
by making the volume of the extracellular compartment large
enough. Such a density effect is observed in glycolytic
oscillations [32].

The alternate mode of chemical communication, which we
assume prevails, is via intercellular junctions that can be
formed upon cell-to-cell contact. These tight junctions can
pass molecules of $\sim 10^3$ MW [33] and therefore, metabolites
and other substances that may exert control over cellular
activities can readily diffuse from cell to cell. Individual
cells in an aggregate can be connected to one or more other
cells and different cells can have a different number of
connections. This leads to a rich variety of topologically
and dynamically distinguishable networks [34], but for
simplicity we regard the entire aggregate as a continuum
contained in a two-dimensional region Ω. When $p\varepsilon[p_0,p_1]$,
one has a continuum of chemical oscillators, linearly
coupled by diffusion.

The governing equation for the system is

$$\frac{\partial c}{\partial t} = D\Delta c + R(c)$$

$$c(\underset{\sim}{r},0) = c_0(\underset{\sim}{r})$$
$$\underset{\sim}{n}\cdot\nabla c = 0$$

in Ω (2)

on $\partial\Omega$.

The diffusivities are positive constants and D is diagonal.
Δ is the Laplacian for the domain Ω and $\underset{\sim}{n}$ is the unit
outward normal. The fact that D is positive definite,
combined with the hypotheses on the kinetics, ensures that
the solution of (2) is componentwise nonnegative for all
time. We assume that it is smooth and bounded pointwise in
space as well, without elaborating the conditions that
guarantee this.

Were $R(c) \equiv 0$, diffusion would always smooth out
initial nonuniformities in concentration and this is the
case in a reacting system as well, provided the diffusivities
are sufficiently large. The following result formalizes
this contention by giving conditions under which the system
always evolves to a uniform state.

THEOREM.   Let

$$\hat{k} \equiv \max_{c} \ ||\frac{\partial R}{\partial c}||$$

and let μ_1 be the smallest non-zero eigenvalue of the
scalar problem

$$\Delta u + \mu u = 0 \qquad \text{in } \Omega$$
$$\underset{\sim}{n} \cdot \nabla u = 0 \qquad \text{on } \partial\Omega.$$

If

$$\min_{i} (D_i)\mu_1 > \hat{k}, \qquad (3)$$

then all spatial nonuniformities decay exponentially in
time.

To derive this result, write the solution of (2) as

$$c(\underset{\sim}{r},t) = \bar{c}(t) + \phi(\underset{\sim}{r},t)$$
$$= Pc(\underset{\sim}{r},t) + Qc(\underset{\sim}{r},t)$$

where $\qquad \displaystyle\int_{\Omega} \phi_i \ d\Omega = 0.$

The operators P and Q≡I-P are projections in the appropriate Hilbert space. Equation (2) can then be written

$$\frac{\partial(\bar{c}+\phi)}{\partial t} = D\Delta(\bar{c}+\phi) + R(\bar{c}+\phi)$$

and by operating on this with P and Q it follows that

$$\frac{\partial\bar{c}}{\partial t} = PD\Delta\phi + PR(\bar{c}+\phi)$$

$$\frac{\partial\phi}{\partial t} = QD\Delta\phi + QR(\bar{c}+\phi).$$

The object is to show that

$$|\phi| \equiv (\int_\Omega ||\phi||^2 \, d\Omega)^{\frac{1}{2}} = (\int_\Omega <\phi,\phi> \, d\Omega)^{\frac{1}{2}}$$

tends to zero as t→∞ when (3) holds. Here and hereafter $||\cdot||$ and $<,>$ denote the Euclidean norm and inner product, respectively.

Analysis of the various terms in the ϕ equation leads to the inequality

$$\frac{d}{dt}|\phi|^2 \leq -\lambda_1|\phi|^2 + \hat{k}|\phi|^2$$

where

$$\lambda_1 \equiv \min_{\substack{|u|=1 \\ \int_\Omega u d\Omega=0}} \int_\Omega <\nabla u, D\nabla u> \, d\Omega.$$

Therefore, if $\lambda_1 > \hat{k}$, $|\phi|^2 \to 0$ exponentially in t. The constant λ_1 is the smallest non-zero eigenvalue of the vector equation

$$D\Delta v + \lambda v = 0$$

$$\underset{\sim}{n} \cdot \nabla u = 0$$

and so

$$\lambda_1 = \min_i (D_i)\mu_1$$

where μ_1 is the smallest non-zero eigenvalue of the scalar equation

$$\Delta u + \mu u = 0$$
$$\underset{\sim}{n} \cdot \nabla u = 0 . \tag{4}$$

This proves the result.

The parameter $\hat{k}$ is a global measure of the <u>sensitivity</u> of the reaction rate to concentration changes. Its reciprocal is the shortest kinetic relaxation time in the system. For some kinetic mechanisms, such as the control mechanisms for inducible or repressible enzymes, estimates of $\hat{k}$ are readily made without knowing <u>a priori</u> bounds on the concentrations. The quantity $[\min_i (D_i)\mu_1]^{-1}$ is the longest relaxation time for diffusion, and if this is less than the shortest kinetic relaxation time, all spatial nonuniformities decay to zero. As a result, (3) is sufficient to ensure that there are no nonuniform steady states. Furthermore, even though these conclusions are derived for zero-flux boundary conditions, they hold for other boundary conditions whenever the eigenvalue problem corresponding to (4) has a zero eigenvalue*.

It should be noted that nothing has been said concerning the nature of the uniform solution. If p is in the range for which (1) has a globally stable steady state, then solutions of (2) always relax to this steady state when (3) holds. When $p\epsilon[p_0,p_1]$, (1) has a globally attracting periodic solution, and the solution of (2) ultimately approaches this solution. The latter case is of interest here because then (3) provides a sufficient condition for a globally-synchronized oscillation, regardless of the initial conditions. A criterion such as this should be of interest in any problem dealing with populations of coupled oscillators. Different modes of coupling will naturally

---

*The dimensionless number $\hat{k}/(\min D_i)\mu_1$ is a Thiele modulus in chemical engineering terminology. For an interesting discussion of the role played by transport limitations in cellular processes see [35].

require a somewhat different analysis and lead to different criteria for synchronization.

To estimate when condition (3) is met, consider a one-dimensional system of length L. The smallest non-zero eigenvalue is $\mu_1 = \pi^2/4L^2$ and a typical value for $\hat{k}$, the pseudo first-order kinetic constant, is 10^{-1} sec^{-1}. Therefore, (3) is satisfied if $\min (D_i) > \sim .04L^2$. If L is 10μ (a typical cell diameter), then the smallest D_i must be larger than $\sim 4.0\times10^{-8}$ cm^2/sec. This is well within the range of diffusivities for small molecules in vitro. If the kinetically active species can freely pass through the junctions, it is to be expected that two cells coupled out of phase will ultimately synchronize. However, when 10 cells are joined in a line, $\min D_i$ must be greater than 4.0×10^{-6} cm^2/sec and this is already near the upper limit of diffusivities in vitro. Of course, the choice of $\hat{k}$ is crucial in these estimates and if the kinetic relaxation time is much longer, as it probably would be in the case of a biochemical oscillation that controls mitosis, the diffusivities can be correspondingly smaller or the lengths correspondingly greater.

IV. LOCAL STABILITY OF PERIODIC SOLUTIONS

The condition that guarantees ultimate synchronization starting from arbitrary initial conditions is a stringent one and can only be met in small systems or when reaction rates are low. At the other extreme in initial conditions, it is clear that a system initially in a uniform periodic state would remain there, but for the inevitable small concentration disturbances. The object here is to find conditions under which these small disturbances decay. The results will partially answer the question of when an artifically-synchronized aggregate of cells will maintain its synchrony. Such questions arise naturally in studies of circadian rhythms.

Let $\Phi(t)$ be the periodic solution of (1) and let T be its period. Write

$$c(\underset{\sim}{r},t) = \Phi(t) + \sum_k y_k(t)u_k(\underset{\sim}{r})$$

where $y_k(t)$ is the amplitude vector for the $k\underline{th}$ eigenfunction $u_k(\underset{\sim}{r})$ of the Laplacian. For small-amplitude disturbances of Φ, (2) can be linearized in y_k and the result is the set of amplitude equations

$$\frac{dy_k}{dt} = (K(t)-\mu_k D)y_k. \tag{5}$$

Here the matrix $K(t)$ is T-periodic: $K(t+T)=K(t)$. The solution for y_k can be written

$$y_k(t) = \Theta_k(t)y_k(0).$$

The matrix $\Theta_k(t)$ is that fundamental matrix of (5) for which $\Theta_k(0)=I$ [36]. Evidently the $\Theta_k(t)$ govern the evolution of the amplitude in the initial disturbance. If $k=0$ corresponds to the zero eigenvalue of the Laplacian, then it follows from the hypotheses on the kinetics that $\Theta_0(T)$ has one eigenvalue equal to one and one eigenvalue less than 1.

The periodic solution $\Phi(t)$ will be an orbitally asymptotically stable solution of (2) if both eigenvalues of $\Theta_k(T)$ are less than one in modulus, or equivalently, if the spectral radius $\rho(\Theta_k(T))$ is less than 1, for all $k \geq 1$. Since $\rho(\Theta_k(T)) \leq ||\Theta_k(T)||$, [37], it suffices to make $||\Theta_k(T)|| < 1$ for $k \geq 1$.

Define

$$\delta = \max_i D_i \qquad\qquad D^* = D-\delta I$$

and write the equation for $\Theta_k(t)$ as

$$\frac{d\Theta_k}{dt} = (K(t)-\mu_k D)\Theta_k = (K(t)-\mu_k\delta I-\mu_k D^*)\Theta_k.$$

Let $\Omega_k(t)$ be defined by

$$\Theta_k(t) = e^{-\mu_k\delta t} \Theta_0(t)\Omega_k(t)$$

and Ω_k will satisfy

$$\frac{d\Omega_k}{dt} = -\mu_k(\Theta_0^{-1}(t)D^*\Theta_0(t))\Omega_k.$$

By converting this to an integral equation and applying Gronwall's inequality [36] to the equation for the norm, it is found that

$$||\Omega_k(T)|| \leq e^{\mu_k \int_0^T ||\Theta_0^{-1}(\tau)D^*\Theta_0(\tau)||d\tau} .$$

Therefore, $||\Theta_k(T)|| < 1$ for $k \geq 1$ if

$$\delta T > \int_0^T ||\Theta_0^{-1}(\tau)D^*\Theta_0(\tau)||d\tau$$

and this in turn is satisfied if

$$\frac{||D^*||}{\delta} < \frac{1}{\max\limits_{t \in [0,T]} \{||\Theta_0^{-1}(t)|| \cdot ||\Theta_0(t)||\}} . \tag{6}$$

Since no use was made in deriving this of the facts that there are only two species and that D is a diagonal matrix, the result is true in general. It is noteworthy that this condition is independent of the eigenvalue μ_k and therefore of geometric factors, in contrast to the condition at (3) for global stability. This is because only ratios of diffusivities enter here. On the other hand, the kinetics enter in a more complicated manner than previously.

The Euclidean norm of the two by two matrix D^* is $|D_2 - D_1|$ and consequently the preceding can be rearranged to read

$$\frac{\chi-1}{\chi} < \frac{D_2}{D_1} < \frac{\chi}{\chi-1} \tag{7}$$

where

$$\chi \equiv \max_{t \in [0,T]} [||\Theta_0^{-1}|| \cdot ||\Theta_0(t)||].$$

χ is never less than 1. The analog of (7) for an n-component system is

$$\frac{\min D_j}{\max D_j} > \frac{\chi - 1}{\chi} .$$

In any event, the conclusion is that the uniform periodic solution is stable with respect to small disturbances if the diffusivities are not too different.

That the diffusivities are not too different is also a sufficient condition for the absence of 'synergistic' or 'diffusive' instabilities of the uniform steady state [6]. A more precise relation between stability of the steady state and stability of the periodic solution is established by the following result.

THEOREM. Suppose that for $k \geq 1$, the matrix $K - \mu_k D$, corresponding to linearization around the uniform steady state, has only eigenvalues with negative real parts. Then uniform periodic solutions of sufficiently small amplitude are orbitally asymptotically stable.

To prove this, consider the amplitude equations for $k \geq 1$:

$$\frac{dy_k}{dt} = (K(t) - \mu_k D)y_k$$

$$= (K - \mu_k D)y_k + (K(t) - K)y_k,$$

and write the solution as

$$y_k(t) = e^{(K - \mu_k D)t} y_k(0) + \int_0^t e^{(K - \mu_k D)(t-\tau)} (K(\tau) - K)y_k(\tau)d\tau.$$

By hypothesis, $K - \mu_k D$ has only eigenvalues with negative real parts so we can find positive constants γ_k and Ω_k such that

$$||e^{(K - \mu_k D)t}|| \leq \Omega_k e^{-\gamma_k t}.$$

It follows that

$$||y_k(T)|| \leq e^{(\Omega_k \tau - \gamma_k)T} ||y_k(0)||$$

where

$$\tau \equiv \max_{[0,T]} \quad ||K(t)-K||.$$

The conclusion follows provided

$$\tau < \min_{k \geq 1} (\frac{\gamma_k}{\Omega_k})$$

and, because $\tau \to 0$ with the amplitude of the periodic
solution, this will be satisfied if the amplitude is
sufficiently small. This is certainly true sufficiently near
the bifurcation points p_0 and p_1.

The correspondence between stability of the steady state
and stability of the periodic solution, both with respect to
nonuniform disturbances, naturally breaks down for large
solutions. Nonetheless, one can sometimes still check
stability of the periodic solution without computing the
Floquet multipliers, as the following result shows.

THEOREM. Let Γ be an annular neighborhood of the periodic
solution and suppose that the matrix $Z \equiv \frac{1}{2}(K+K^T) - \mu_k D$ has only
negative eigenvalues for $c \in \Gamma$. Then $\Phi(t)$ is orbitally
asymptotically stable.

This follows directly from the estimate [37]
$||y_k(t)|| \leq ||y_k(0)|| \exp\{\int_0^t \lambda_{max}^Z d\tau\}$. To apply it, one
needs an estimate of the location of $\Phi(t)$. This is
sometimes easy to obtain, particularly for relaxation
oscillations.

V. SECONDARY BIFURCATIONS OF NONUNIFORM PERIODIC SOLUTIONS

Aside from the special case in which the diffusivities
are equal and diffusion can never lead to destabilization of
a uniform periodic solution, it is impossible to analyze
destabilization of periodic solutions in the generality
possible for steady states. However, one result of the
preceding section is that stability of small-amplitude
periodic solutions goes hand in hand with stability of the

steady state with respect to nonuniform disturbances, and
therefore the latter question should be addressed when p
is near a bifurcation point. This requires analysis of the
eigenvalues of the pencil of matrices $K-\mu_k D$, $k \geq 1$. When
K and D are 2×2 these eigenvalues are

$$\lambda_\pm = \frac{1}{2}\ (TrK - \mu_k TrD \pm \sqrt{\Delta(\mu_k)})$$

where the discriminant Δ is given by
$$\Delta(\mu_k) = (D_2 - D_1)^2 \mu_k^2 + 2(k_{11} - k_{22})(D_2 - D_1)\mu_k + (k_{11} - k_{22})^2 + 4k_{12}k_{21}$$
For the present we take p near p_0 or p_1 and therefore
can assume that the constant term in Δ is negative. It
follows that the eigenvalues are always complex for small
μ_k and real for large μ_k. If both diffusivities are
strictly positive the eigenvalues are negative for large μ_k
and there is at most a finite number of μ_k's for which
either eigenvalue has a positive real part.

The following discussion will be simplified if we
restrict attention to a system in which there are only two
Fourier modes. Therefore we drop the continuum description
temporarily and focus on a system of just two coupled cells.
In this case one mode is uniform and the other is
nonuniform; these have eigenvalue 0 and 2, respectively [34].
Moreover, we shall assume that $\Delta(2) > 0$, as the other case
is uninteresting. As a result, there are only five distinct
types of λ vs μ_k diagrams in which at least one eigenvalue
has a zero real part, as shown in Figure 1.

Which one of these obtains depends on the relationship
between the parameters p, D_1 and D_2. Certainly $p = p_0$
or $p = p_1$ in (b), (c) and (e) because bifurcation of
uniform periodic solutions is independent of D_1 and D_2.
The loci along which there is a zero real eigenvalue
at $\mu_k = 2$ are given by

$$J \equiv \det(K - 2D) = 4D_1 D_2 - 2(k_{11}D_2 + k_{22}D_1) + \det K = 0.$$

The zero eigenvalue is simple, and therefore, in every case
except possibly (e), a nonuniform steady state bifurcates
from the uniform steady state when parameters cross the

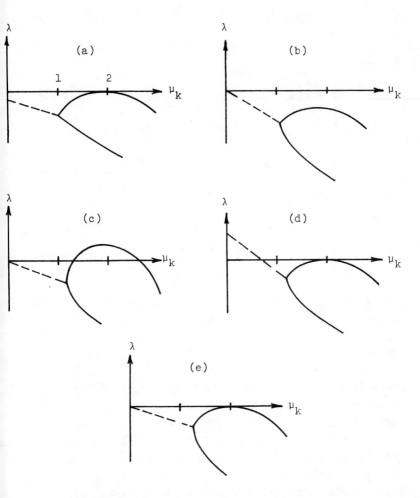

igure 1. Eigenvalue vs μ_k diagrams when one eigenvalue
has a zero real part. ----: Real part of a complex
eigenvalue, —— real eigenvalue. Only the values at
$\mu_k=0$ and 2 have meaning in the present context.

ocus $J=0$. At any fixed p the kinetic coefficients are
ixed and this locus is a hyperbola in the D_1-D_2 plane, as
hown in Figure 2. In the continuum case there is a countable
umber of such curves, the $k^{\underline{th}}$ having D_1-intercept
et $K/\mu_k k_{22}$ and horizontal asymptote k_{22}/μ_k. How these

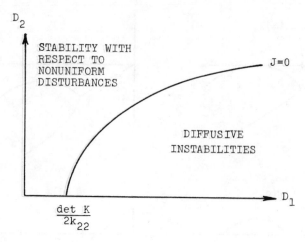

Figure 2

curves vary with p depends on det K and k_{22}; if for
instance det $K/\mu_k k_{22}$ has a single turning point for
$p\epsilon[\hat{p}_0,\hat{p}_1]$ then we have the following picture for a fixed
non-zero μ_k. When there are a finite or countable number
of positive μ_k there is a second family of loci,
parameterized by μ_k. The ZZKK mechanism [29] for the
Zhabotinskii-Belousov reaction leads to a diagram like that
shown.

A Hopf bifurcation of a uniform periodic solution always
occurs upon crossing the lines $p=p_0$ and $p=p_1$, and by
hypothesis the periodic solution exists only for $p\epsilon[p_0,p_1]$
and is stable with respect to uniform disturbances.
Diffusion does not affect these bifurcations when zero-flux
boundary conditions are imposed. Now suppose that D_2 is
fixed and let D_1^* be the intersection of the corresponding
$J=0$ locus with the line $p=p_0$ (Figure 3). If $D_1>D_1^*$,
then as p increases, the uniform steady state loses
stability with respect to the nonuniform disturbance at some
$\tilde{p}<p_0$. At this p-value the applicable $\lambda-\mu_k$ diagram is
Figure 1(a) and a nonuniform steady state bifurcates from
the uniform steady state. One can show that with zero-flux
(or periodic) boundary conditions the bifurcating solution

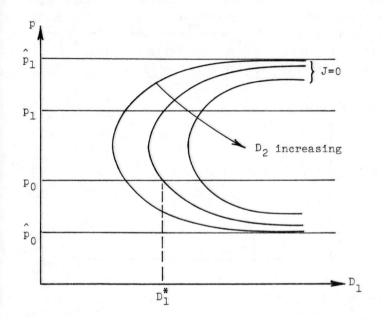

Figure 3. The loci J=0 for variable D_2. The lines $p=\hat{p}_0$ and $p=\hat{p}_1$ are the loci $k_{22}=0$; $\hat{p}_0=0$ and $\hat{p}_1=\infty$ are allowed.

xists on only one side of the bifurcation point. Let us uppose that for (D_1,p) near (D_1^*,p_0) the bifurcation is upercritical; then the bifurcating solution is stable. At $=p_0$ the periodic solution emerges and at this point the ppropriate $\lambda-\mu_k$ diagram is Figure 1(c). Note that the eriodic solution is <u>unstable</u> for p near p_0, because the onuniform mode grows in time.

If $D_1-D_1^*$ is small and positive, the amplitude of the onuniform solution is small when p is near p_0 and the inearization of the nonlinear equations along the ifurcating branch has a pair of complex conjugate eigen- alues near those of K. By making $D_1-D_1^*$ sufficiently mall, one can ensure that there is a point p* near p_0 t which this complex pair of eigenvalues has a zero real

part. At this point a secondary bifurcation occurs and a
stable nonuniform periodic solution bifurcates supercriticall
If it happens that $p^*>p_0$, the amplitude vs p diagram is
as shown in Figure 4. Some computational results that
correspond to this case are given in [38].

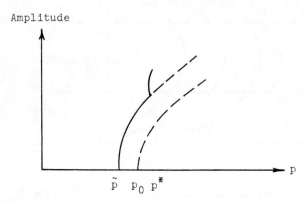

Amplitude

$\tilde{p}$ p_0 p^*

p

Figure 4. The amplitude vs p diagram when the uniform
  solution first becomes unstable to nonuniform disturbances
  Solid lines: stable solutions;dashed lines: unstable
  solutions. The mirror image steady state, in which the
  cells are interchanged, is omitted.

    Let us summarize the preceding. When $\tilde{p}$-p is small and
positive, the linearization around the uniform solution has
two real negative eigenvalues and a complex conjugate pair
with negative real part. As p crosses $\tilde{p}$, one real
eigenvalue crosses the imaginary axis and a nonuniform
solution emerges, while at $p=p_0$, the real part of the
complex pair crosses the imaginary axis and a uniform
periodic solution emerges. At $\hat{p}$, a nonuniform periodic
solution emerges from the nonuniform steady state. A
qualitative understanding of the corresponding changes in the
phase portrait can be gotten from the following three-
dimensional section of the four-dimensional phase space.
This cross-section omits the direction corresponding to the
real eigenvalue that always remains negative. Distance
along the vertical axis corresponds to the amplitude of the

onuniform mode.

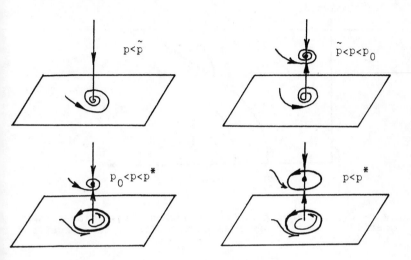

igure 5. A three-dimensional section of phase space.
The mirror image steady state is omitted.

The order in which the bifurcations occur is reversed
rom the preceding when $D_1-D_1^*$ is small and negative. The
niform periodic solution is now stable and the bifurcating
onuniform steady state is unstable. It is to be expected
hat if $D_1-D_1^*$ is small enough there will be a point on the
ranch of uniform periodic solutions at which an eigenvalue
f $\theta_1(T)$ crosses the unit circle. It can be shown by
sing Abel's formula [36] that in a two-component reacting
ystem with a stable periodic solution, the eigenvalues of
$\theta_k(T)$, $k\geq1$, can only cross the unit circle at ±1.
herefore, when bifurcation of a new periodic solution occurs,
t must have period close to T or $2T$ near the
ifurcation point. Now the amplitude <u>vs</u> p diagram is as
hown in Figure 6.

This picture has been confirmed by numerical integration
f the equations for the ZZKK mechanism; the details will
e reported elsewhere.

80   H. G. OTHMER

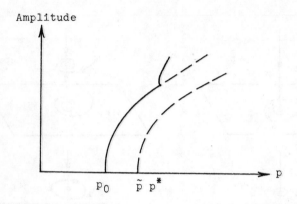

Figure 6. The analog of Figure 4 when bifurcation of the
   periodic solution precedes bifurcation of the nonuniform
   steady state.

As D_1 varies in a neighborhood of D_1^*, a curve is
generated in the $D_1 p$-plane along which the nonuniform
periodic solutions emerge. As D_2 varies, this curve
generates a bifurcation surface. As a result, the nonuniform solutions exist in some open set of parameter values.
In the continuum case there is the further possibility that
the nonuniform periodic solutions become unstable and a
tertiary branching of another periodic solution occurs.
The complete picture has not been worked out for even the
simplest case of one nonuniform mode, but it should be
evident by now that the kind of destabilization of a uniform
periodic solution that we have described is a robust
phenomenon that must be considered whenever the self-
activating species is also the slower diffusing species.

VI. CONCLUSION
   It is interesting, in light of the experimental evidence
linking the absence of intercellular communication with
abnormal growth [26], that the asynchronous states found
here exist only when the diffusivities are sufficiently

ifferent. As intercellular exchange of the self-activating
pecies increases, the spatially nonuniform solutions
isappear, leaving only the synchronous oscillation. One
s tempted to identify the 'resting' state of the cellular
etwork with this uniform state and mitotic activity with
 nonuniform state, in direct contrast to Burton and Canham's
nterpretation. In this vein, it is noteworthy that the
niform periodic solution only exists when there is no flux
cross the boundary of the network or when the network is
ne unit is a repeated pattern. Interruption of inter-
ellular communication, for example by wounding, destroys
he uniform state and might thereby stimulate cell division.

    The nonuniform periodic solutions we have described
epresent standing oscillations, in contrast to the
ropagating waves that are found in infinite domains [39].
n the two-cell case studied here, the cells are 180° out of
hase in the nonuniform mode, but more complicated spatial
atterns of phase relationships can arise when more cells
re present. These patterns provide the time-dependent
nalog of the stationary morphogentic 'maps' discussed in
he Introduction. How nonuniform patterns of phase can be
ranslated into a spatio-temporal framework for gene
ctivation is discussed elsewhere [40].

    It has been assumed throughout that all parameters in
he chemical kinetics are spatially uniform and as a result,
he period of oscillation is uniform throughout the system.
ecause of cell-to-cell variations in enzyme and substrate
oncentrations, this will not be true in reality. Nonethe-
ess, if the spatial variation in the period is small one
ay still expect to find a spatially-synchronized
scillation under conditions similar to those given here.
his has been established for a two-cell system [41] but
pparently not for a continuum description of a multicellular
ystem. Recent experiments using the Belousov-Zhabotinskii
eaction in coupled stirred reactors illustrate
ynchronization at a common frequency for small frequency
ifferences and subharmonic synchronization for large
requency differences [42].

REFERENCES

1. L. Wolpert, 'Positional Information and the Spatial
        Pattern of Cellular Differentiation', J. Theor.
        Biol., 25, 1 (1969).

2. A.M. Turing, 'The Chemical Basis of Morphogenesis',
        Phil. Trans. Roy. Soc., B237, 37 (1952).

3. J. Maynard-Smith and K.C. Sondhi, 'The Arrangement of
        Bristles in Drosophila', J. Embry. Exp. Morph.,
        9, 661 (1961).

4. J.I. Gmitro and L.E. Scriven, 'A Physicochemical
        Basis for Pattern and Rhythm', in Intracellular
        Transport (K.B. Warren, ed.) Academic Press (1966)

5. A.N. Zaiken and A.M. Zhabotinsky, 'Concentration Wave
        Propagation in Two-dimensional Liquid-phase
        Self-oscillating System', Nature, 225, 535 (1970).

    A.T. Winfree, 'Spiral Waves of Chemical Activity',
        Science, 175, 634 (1972).

    J.A. DeSimone, D.L. Beil and L.E. Scriven, 'Ferroin-
        Colloidion Membranes: Dynamic Concentration
        Patterns in Planar Membranes', Science, 180, 946
        (1973).

    N. Kopell and L.N. Howard, 'Pattern Formation in the
        Belousov Reaction', Lects. on Math. in the Life
        Sciences, 7, 201, (1974).

6. H.G. Othmer and L.E. Scriven, 'Interactions of Reaction
        and Diffusion in Open Systems', Ind. Eng. Chem.
        Fund. 8, 302 (1969).

    L.A. Segel and J.L. Jackson, 'Dissipative Structure:
        An Explanation and an Ecological Example',
        J. Theor. Biol. 37, 545, (1972).

7. C. Georgakis and R.L. Sani, 'On the Stability of the
        Steady State in Systems of Coupled Diffusion and
        Reaction', Arch. Rat. Mech. Anal., 52, 266 (1973).

8. J.F.G. Auchmuty and G. Nicolis, 'Bifurcation Analysis of
        Nonlinear Reaction - Diffusion Equations', Bull.
        Math. Biology, 37, 323, (1975).

9. A. Gierer and H. Meinhardt, 'A Theory of Biological
        Pattern Formation', Kybernetik, 12, 30 (1972).

10. S.A. Levin, 'Spatial Patterning and the Structure of
        Ecological Communities', Pages 1-36 in S.A. Levin,
        ed. Lectures on Mathematics in the Life Sciences,
        Vol. 8: Some Mathematical Questions in Biology
        VII (1976).

    L.A. Segel and S.A. Levin, 'Application of Nonlinear
        Stability Theory to the Study of the Effect of
        Diffusion on Predator-Prey Interactions', in
        Topics in Statistical Mechanics and Biophysics:
        A memorial to Julius L. Jackson, AIP Conference
        Proceedings, No. 27 (in press).

1. T.J. Mahar and B.J. Matkowsky, 'A Model Biochemical
        Reaction Exhibiting Secondary Bifurcation',
        preprint.

2. H.G. Othmer and L.E. Scriven, 'Nonlinear Aspects of
        Dynamic Pattern in Cellular Networks', J. Theor.
        Biol., 43, 87 (1974).

3. H.G. Othmer, 'Nonlinear Wave Propagation in Reacting
        Systems', J. Math. Biology, 2, 133 (1975).

4. E.K. Pye, 'Biochemical Mechanisms Underlying the
        Metabolic Oscillations in Yeast', Can. Jour. Bot.,
        47, 271 (1969).

5. T. Pavlidis and W. Kauzmann, 'Toward a Quantitative
        Biochemical Model for Circadian Oscillators',
        Arch. Biochem. Biophys., 132, 338 (1969).

    E.K. Pye, 'Periodicities in Intermediary Metabolism',
        in Biochronometry, Friday Harbor Symposium, 623
        (1971).

6. A.T. Winfree, 'Biological Rhythms and the Behavior of
        Populations of Coupled Oscillators', J. Theor.
        Biol., 16, 15 (1967).

    T. Pavlidis, 'Populations of Biochemical Oscillators as
        Circadian Clocks', J. Theor. Biol., 33, 319 (1971).

7. A. Westerfeld and M.A. Freeke, 'Cell Cycle of Multi-
        nucleate Cells after Cell Fusion', Exp. Cell Res.,
        65, 140 (1971).

8. H.P. Rusch, W. Sachsenmaier, K. Behrens, and V. Gruter,
        'Synchronization of Mitosis by the Fusion of the
        Plasmodia of Physarum polycephalum', J. Cell.
        Biol., 31, 204 (1966).

9. K. Esser, 'Genetic and Biochemical Analysis of
        Rhythmically Growing Mycelia in the Ascomycete
        Podospora anserina', J. Int. Cycle Res., 3, 123
        (1972).

0. G. Kraepelin and G. Franck, 'Self-synchronization in
        Yeast and Other Fungi', Int. Jour. Chronobiol.,
        1, 163 (1973).

1. E. Guttes, S. Guttes and H.P. Rusch, 'Morphological
        Observations on Growth and Differentiation of
        Physarum polycephalum Grown in Pure Culture',
        Dev. Biol., 3, 588 (1961).

2. T. Pavlidis, Biological Oscillators: Their Mathematical
        Analysis, Academic Press (1973).

3. C.H. Waddington, 'Autogenous Cellular Periodicities
        as (a) Temporal Templates and (b) the Basis of
        "Morphogentic Field"', J. Theor. Biol., 8, 367
        (1965).

4. B. Goodwin and M. Cohen, 'A Phase-shift Model for the
        Spatial and Temporal Organization of Developing
        Systems', J. Theor. Biol., 25, 49 (1969).

25. A.C. Burton and P.B. Canham, 'The Behavior of Coupled
     Biochemical Oscillators as a Model of Contact
     Inhibition of Cellular Division', J. Theor. Biol.
     39, 555 (1973).

26. W.R. Loewenstein and Y. Kanno, 'Intercellular
     Communication and the Control of Tissue Growth.
     Lack of Communication between Cancer Cells',
     Nature, 209, 1248 (1966).

27. S. Kauffman and J.J. Wille, 'The Mitotic Oscillator in
     Physarum polycephalum', J. Theor. Biol., 55, 47
     (1975).

28. D. Ruelle and F. Takens, 'On the Nature of Turbulence'
     Comm. Math. Phys., 20, 167 (1971).

29. A.M. Zhabotinskii, A.N. Zaiken, M.D. Korzukhin and
     G.P. Kreitser, 'Mathematical Model of a Self-
     oscillating Chemical Reaction', Kinet. Cat., 12,
     516 (1971).

30. H.G. Othmer, 'On the Temporal Characteristics of a
     Model for the Zhabotinskii-Belousov Reaction',
     Math. Biosci, 24, 205 (1975).

31. J. Aldrige and H.G. Othmer, in preparation.

32. J. Aldridge and E.K. Pye, 'Cell Density Dependence of
     Oscillatory Metabolism', Nature, 259, 670 (1976).

33. W.R. Loewenstein, 'Emergence of Order in Tissues and
     Organs', Dev. Biol. Supp., 2, 151 (1968).

34. H.G. Othmer and L.E. Scriven, 'Instability and Dynamic
     Pattern in Cellular Networks', J. Theor. Biol.,
     32, 507 (1971).

35. P. Weisz, 'Diffusion and Chemical Transformation',
     Science, 179, 433 (1973).

36. F. Brauer and J. Nohel, Qualitative Theory of Ordinary
     Differential Equations, W.A. Benjamin, Inc. (1969)

37. J.L. Daleckii and M.G. Krein, Stability of Solutions
     of Differential Equations in Banach Space, Am.
     Math. Soc. (1974).

38. J. Tyson and S. Kauffman, 'Control of Mitosis by a
     Continuous Biochemical Oscillation', J. Math.
     Biol., 1, 289 (1975).

39. N. Kopell and L.N. Howard, 'Plane Wave Solutions to
     Reaction-Diffusion Equations', Stud. Appld. Math.,
     52, 291, 1973.

     P. Ortoleva and J. Ross, 'On a Variety of Wave Phenomen
     in Chemical Reactions', J. Chem. Phys., 60,
     5090, 1974.

40. Z. Hejnowicz, 'A Model for Morphogenetic Map and Clock'
     J. Theor. Biol., 54, 345 (1975).

41. V. Torre, 'Synchronization of Non-linear Biochemical
     Oscillators Coupled by Diffusion', Biol. Cyber.,
     17, 137 (1975).

2.   M. Marek and E. Svobokova,   'Nonlinear Phenomena in
          Oscillatory Systems of Homogeneous Reactions-
          Experimental Observations', Biophys. Chem., $\underline{3}$,
          263 (1975).

    M. Marek and I. Stuchl,   'Synchronization in Two
          Interacting Oscillatory Systems', Biophys. Chem.
          $\underline{3}$, 241 (1975).

Department of Mathematics
Rutgers University
New Brunswick, New Jersey

ectures on Mathematics in the Life Sciences
olume 9, 1977

DYNAMIC MODELS OF THE MITOTIC CYCLE:

EVIDENCE FOR A LIMIT CYCLE OSCILLATOR[*]

Stuart Kauffman, M.D.
Associate Professor
Department of Biochemistry and Biophysics
School of Medicine
University of Pennsylvania
Philadelphia, PA

ACKNOWLEDGEMENTS

This work was a completely collaborative effort, most notably with
Drs. John Wille and John Tyson, and Mr. Carl Sheffy, at the University of
Chicago. I am deeply grateful to all three for joint experimental and
mathematical work. The work was partially supported by NSF grant GB36067.

---

*This is a revised version of an article entitled "Mathematical Models
in Biological Discovery", Lecture Notes in Biomathematics (D.L. Soloman and
C.F. Walter, eds.), Vol. 13, pp. 95-131, published by Springer-Verlag,
Heidelberg reprinted with permission of the publishers.

INTRODUCTION: Models of the Mitotic Cycle

Control of the periodicity of mitosis, and maintenance of stable phase relation among events of the cell cycle, have been extensively studied in a number of prokaryocytes (1,2) in tetrahymena (3,4) Paramecium (5), yeast (6), Physarum (7,8) and higher cells (9). Two broad classes of models have emerged over the past several decades, (reviewed by J.M. Mitchison 10). The first envisions a recurrent sequence of discrete cellular events, related to one another as a simple causal loop sequence, or as partially independent, partially connected causal sequences eventually forming a loop. Hartwell's (11) elegant genetic analysis of the yeast division cycle using temperature sensitive mutants stopping cells as specific phases has led him to this type of model. Such a causal loop model would explain both the periodicity of the cycle and the maintenance of proper phase relations between events of the cycle. The events of the cycle are here considered as parts of the clock.

The second broad class of models proposes the continuous accumulation of a mitogen, or division protein during the cell cycle which reaches a critical concentration, perhaps converts to a new division structure, initiates mitosis, is used up in mitosis, and reaccumulated during the next cycle. This is a form of a central clock model in which mitosis is part of the clock, by causing mitogen to fall, but other events of the cycle are driven by mitosis, by one another, or by the clock. Zeuthen (3,9) and his colleagues (12) have performed extensive heat shock experiments on tetrahymena which led to such a model.

The experiments we have performed on the syncytial plasmodium of the myxomycete Physarum polycephalum during the past two years, have convinced us that these two theories, which have dominated accounts of the cell

ycle, are inadequate. Our experiments have induced us to try
o formulate a more adequate theory, based on the properties of sustained
iochemical oscillations, to account for cellular mitotic periodicity, and we
ave already subjected our model to severe tests.

ausal Loop Model

The view of the cell cycle as a sequence of distinct events standing
n causal relations to one another, and eventually forming a loop leads
aturally to the following questions: what events occur at each phase of
he cycle, and what are the causal relations among these events. Classical
ork based on this view established the length and phase of the DNA synthesis,
, period in a variety of organisms, and allowed the definition of the G_1
eriod preceding S, and G_2 following S and terminating in mitosis. We do
ot discuss this classical work further except to note that in Physarum,
s in other lower eukaryotes, no G_1 period exists. The S, G_2, M, G_1 sequence
eparates the cell cycle into three phase zones. Far finer discrimination
of phase markers has become available by analysing the specific time during
the cycle when certain enzyme activities rapidly increase in a step which
is thereafter maintained, or briefly peak, then return to a basal level.
In Physarum, about nine phase specific step and peak enzymes are known
(13, 14, 15, 47), in Saccharomyces cerevisiae at least 22 are known (16, 17,
18, 19,20,21,22) in mouse L cells seven step and peak enzymes are known
(23,24,25,26,27) and so on. The existence of such phase markers implies
some kind of cellular clock, but leaves unclear whether these enzyme markers
are merely "hands" of the clock, driven by some more central mechanism,
or are themselves parts of the mechanism underlying the periodicity.

One popular theory to account for the maintenance of phase relations
among enzyme activities postulates a linear reading (transcribing) of genes
along the chromosomes, such that spatial sequence dictates temporal sequence.
Tauro et al. (1968) have compared the step timings of nine enzymes with the

positions of their genes in S. cerevisiae and found them consistent with an
end to end reading along the chromosomes. Of these nine, at least four are
located on the same chromosome. This interesting possibility would account
for maintenance of phase relations among markers on any chromosome, but not
for coordination between enzymes on different chromosomes, nor for periodicity
in enzyme synthesis.

Transition Points in the Cycle

The closed causal loop model leads naturally to search for transition
points between various events, or stages of the mitotic cycle. The earlier
experiments in this area were accomplished with diverse metabolic inhibitors
or temperature shocks. Thus, a temperature shock applied to Physarum plasmodia
60 minutes before mitosis delays the subsequent mitosis several hours (28),
but the same shock applied within 7 minutes of metaphase does not delay
metaphase or the subsequent completion of the mitotic sequence and entry into
S. Using actinomycin D, puromycin, and X-ray, Doida and Okada (29) have been
able to establish a number of transition points in mouse L5178Y cells:
switching off of DNA synthesis, switching off of nuclear RNA synthesis
switching on of nuclear RNA synthesis. Muldoon et al. (30) showed in
Physarum, using cycloheximide, that DNA replication itself is divisible
into ten distinct rounds, in which de novo protein synthesis at ten distinct
times in S is required to replicate another quantitized fraction of the nuclear
DNA.

The most elegant analysis based on the closed causal loop model has
been carried out by L. Hartwell and his colleagues (11). They reason that
the mitotic cycle can best be dissected into its distinct events by collecting
conditional mutants, each blocking the cell at a specific phase of the cycle.
The enormous virtues of this approach are that it allows a far finer discrimination of events in the cycle than can be revealed by defining transition
points with respect to broad classes of biochemicals such as DNA, RNA or

-otein synthesis; and that there is a significant hope of establishing
ependent pathways in which early events in the cycle can be shown to be
ecessary (but not sufficient) conditions for the occurrence of certain
ibsequent events of the cycle.

Hartwell's analysis (11) has led him to define such a model for the
east division cycle. His temperature sensitive mutants define two apparent
iusal sequences in which a block at an earlier step blocks all subsequent
vents of the sequence: 1. a "start" event, initiation of DNA synthesis,
intinuation of DNA synthesis, medial nuclear division, late nuclear division
id the "start" event. 2. start, bud emergence, bud elongation, nuclear
gration. Hartwell proposed that the two sequences must join to cause
vtokinesis and cell division, but that the first sequence alone suffices
) close the causal loop to initiate the start event.

Hartwell's analysis is currently the most thorough mutant analysis of a
vision cycle. He has strongly favored a closed causal loop model of the
ell cycle. Nevertheless, he has two mutants which cast doubt on the simple
ew presented. One mutant blocks the initiation of DNA synthesis, an event
iken to lie on the closed causal loop. Despite this, the organism goes
irough repeated rhythmic rounds of bud emergence. Thus, closed sequence 1
not a necessary condition for rhythmic bud emergence. Similarly, he has
mutant blocking bud emergence which goes through at least two rounds of S,
ince the second sequence is not necessary for the first. This data can
till be accommodated within a closed causal loop model; one merely postulates
vo separate closed causal loops, one involving DNA synthesis, the second
ivolving bud emergence, and postulates that these are normally held in
»propriate phase relations to one another. For example, budding rhythmically
ight be clocked by bud size, increasing to a threshold before initiation of a
2w bud event. An alternate interpretation of this data is that there is some
eparate central cellular clock which ticks off "start" events and thereby

drives the two separable causal sequences which are "downstream" of the central clock. Hartwell himself appears to favor a hybrid model, in which the DNA-start sequence is a closed causal loop, while bud emergence may be timed by some separate, perhaps central, clock.

Despite its enormous power, the use of conditional mutants blocking the cell at specific phases of the cycle suffers serious limitations in any attempt to account fully for the cell cycle. The issue can be drawn by an analogy with a grandfather clock. The weights provide the energy to drive the mechanism. The gears and escapements marshall the energy into quantitized changes occurring in the proper repeating phase sequence, the hands show the hour, and the pendulum guarantees the time periodicity. Without the pendulum, there may be a closed sequence of events which can cycle, but intervals of time are not counted, and there is no clock. We will propose below that the cells contain an entity analogous to a pendulum, a sustained biochemical oscillation, which provides the timing of events. It is critically important that a pendulum, and a biochemical oscillation, manage to behave periodically due to their continuous dynamical laws of motion, not by virtue of a closed sequence of discrete states. In terms of this analogy, mutants of a grandfather clock stopping it at specific phases will reveal the machinery of the hands, and of the gears and escapements, but not of the central pendulum. Thus, a critical limitation of conditional mutants blocking the cell at specific phases is that, by picking out such phases specific events, it inherently overlooks continuous dynamical processes which may be involved in timing, such as possible sustained biochemical oscillations. We next consider the substantial evidence for such continuous timing processes.

Mitogen Models

The second broad class of models to account for the periodicity of mitosis supposes the accumulation of a mitogen during the cycle which triggers mitosis at a critical concentration, is used up in the process, and must reaccumulate the next cycle. Good evidence supports the hypothesis of a cytoplasmic

"itogenic" stimulus, Most binucleate cells achieved by somatic cell fusion

ow mitotic synchrony (31, 32) and syncytial organisms such as Physarum show

artling synchrony of nuclear mitosis. In Physarum, up to 10^{10} nuclei in a

mmon cytoplasm divide within two minutes of one another in a ten hour cycle.

rong evidence for the accumulation of a mitogen during the cycle was provided

en it was shown that fusion of a Physarum plasmodium due to undergo mitosis

1 pm with a second plasmodium due to undergo mitosis at 3 pm led to their

ncronous mitosis at 2 pm. If considerably more of the 1 pm plasmodium were

ed than the 3 pm plasmodium, the fused pair underwent mitosis at 1:30. Physarum

rikingly demonstrates that phase, measured as time to synchronous mitosis,

haves as though it can be averaged by mixing a 1 pm and 3 pm cytoplasm. This

rongly suggests that fusion manages to average phases by averaging the concen-

ations of one or more continuously changing biochemical variables. On the

mplest account, the 1 pm plasmodium has a higher concentration of mitogen

an the 3 pm plasmodium, and on fusion, the concentration averages out to the

pm level. This data supplies the best evidence that phase behaves as though

were specified by the levels of one or more continuously graded variables

ke biochemical concentrations.

Weaker evidence that phase is specified by one or more substances whose

ncentrations vary continuously during the cell cycle is provided by experiments

Zeuthen and coworkers (3,4,12) using heat shocks on tetrahymena. They found

at a heat shock early in the cycle does not delay the subsequent mitosis. As

e cycle progresses, the same shock produces a successively longer mitotic

lay, until late G_2 when the delay declines rapidly to nothing. The same form

delay curves induced by heat shocks has been found in a variety of organ-

ms including Physarum (28) and higher cells (9). Such delay curves have

iformly been interpreted to indicate the absence of mitogen synthesis early

the cycle, the gradual accumulation of mitogen during the cycle, such that

heat shocks later produce longer delays, then the conversion of the heat labi
mitogen to a heat stable "division structure" shortly before mitosis, to acco
for the drop in delay for shocks applied shortly before mitosis. We will
provide an alternate interpretation of these results below. However, we wou?
agree that the gradual increase in mitotic delay for the same heat shock app?
successively later in the cycle suggests the continual increase in concentra?
of some phase specifying substance, or substances, subjected to first order
destruction in the heat shock environment.

Thus cells give clear evidence both of discrete events occurring at speci
phases of the cycle, which stand in casual relations to one another and also
give evidence that phase can be averaged as though it were specified by a
continuously graded "concentration" of one or more substances. The simple c
loop causal model has a difficult time accounting for the phase averaging
phenomena. The simplest mitogen model, with accumulation of a single variab?
to a threshold, is both unlikely biochemically, and unable to account for the
discrete events happily. A more adequate general conceptual framework is re?

Biochemical Oscillations.

A start towards a more adequate framework is supplied by the large body
of work done on circadian rhythms. A compendious review is given by Pavlidis
(33). The conceptual apparatus which has gradually come to be employed in
this field utilizes the concept of sustained oscillations in complex, non-li?
dynamic systems. The cellular basis for such circadian rhythms are unestabl?
and in cases such as the eclosion rhythm in Drosophila, hard to imagine. Bu?
the similarity in phase resetting phenomena after heat shocks in many cell
systems, leads naturally to the hypothesis that they are underlain by simila?
general mechanisms.

The possibility that biochemical oscillations might play a role in
governing the periodicity of mitosis is strongly enhanced by the recent disc?
very that such sustained biochemical oscillations do occur in cells. The cl?

est example is the glycolytic oscillation in yeast, in which that pathway can exhibit sustained spontaneous oscillations of pathway constituents, due directly to the kinetic equations linking the synthesis and degraduations of the biochemicals (34,35). Spontaneous biochemical oscillations seem to afford one natural kind of cellular clock, and several authors have suggested such oscillations might control mitosis (36,37). Our own simple model proposes an oscillation of two biochemicals, in which a threshold level in one triggers mitosis.

Is mitosis part of the mitotic clock?

A natural question which arises given any of the views of the mitotic cycle we have described, is whether specific cell events such as mitosis, DNA synthesis, or the occurrence of particular step or peak enzymes which serve as phase markers are part of the clock, or are causally downstream of the clock. In the simple causal loop model, an event A is part of the clock if it is a member of the closed causal loop generating periodicity, such that blocking A stops the clock, while cells accumulate at the blocked phase. If other phase markers continue to cycle when event A is blocked, then A is not considered part of the clock. Bud emergence is an example of the latter, as described by Hartwell (11). A mutant blocking bud emergence does not block repeated rounds of DNA synthesis.

Hartwell has shown in yeast a number of temperature sensitive mutants which do block division at specific phases, and thus accumulate cells at that phase. This has been taken as evidence that such events are parts of the clock in the sense that the mutants block the casual loop and stop the clock. However, the experiments performed do not prove that the cells' division clock has stopped at all, they may merely prove that _expression_ of any underlying persisting rhythmicity is blocked. This is a critical point. Hartwell's example of a mutant blocking initiation of the DNA replication which nevertheless undergoes repeated rhythmic budding events strongly suggests an underlying rhythmic process which continues despite the block in DNA synthesis, and triggers his

"start" events. Even if, in the non-permissive temperature conditions, all cell
collect at some morphological stage of the cycle, and underlying clock may still
be ticking independently. To our knowledge, there is at present no convincing
demonstration that any mutant blocking cells at a specific phase actually stops
the cell clock, rather that its hands.

In a series of interesting experiments, Mano (38,39) has obtained striking
evidence that the mitotic clock may not only be independent of mitosis, but in
some cases independent of DNA synthesis, and RNA synthesis as well. He analysed
a regular cyclic variation in protein synthesis in a cell free system derived
from the supernate of a homogenate of cleavage stage sea urchin embryos. The
rhythm persisted in the presence of actinomycin. The rhythm is not present
in unfertilized eggs. ATP, GTP and a maternal mRNA derived from fertilized
eggs acted as developers of the intrinsic protein synthesis rhythm in super-
nates of unfertilized eggs, but did not determine the phase of that rhythm.
Mano found that the SH content of the KCl soluble protein fraction also varied
cyclically in parallel with cyclic incorporation of amino acids, and may
regulate factors determining a cyclic variation in the binding of aminoacyl-
tRNA to ribosomes, and therefore induce a cyclic variation in amino acid
incorporation. Mano found that histone synthesis continued cyclically in the
presence of sufficient cytosine arabinoside to inhibit DNA synthesis almost
completely. In addition, tubulin was found to be synthesized rhythmically,
without DNA synthesis, but at a different phase than histones.

Mano's experiments not only suggest that mitosis is not a necessary part
of its own clock, but fit well with the general hypothesis that the clock
consists in a sustained biochemical oscillation of many variables, perhaps
acting largely at the level of post transcriptional controls.

Clock variables

Efforts to find the actual clock variables have been less than successful.
In Physarum, Oppenheim and Katzir (40) have reported that treatment of early

$_2$ plasmodia with extracts of late G_2 plasmodia phase advanced the mitosis of
the treated material. Efforts by Sachsenmaier (41), Mohberg (42), and ourselves
to repeat these experiments have not confirmed Oppenheim and Katzir. Attempts
to inject late G_2 material into the veins of Physarum plasmodia have succeeded,
but without effect on subsequent mitosis. Technically, this is understandable,
since small volumes of material can be injected into plasmodia which must be 3
to 5 centimeters in diameter to have adequate veins, and thus represent considerable tissue mass. Recently, Bradbury et. al. (43,44) have reported that
phosphorylation of the F_1 histone fraction precedes chromosomal condensation,
that the activity of the corresponding phosphorylating enzyme peaks in mid G_2,
and that addition of exogenous calf F_1 histone phosphorylating enzyme to the
surface of plasmodia in mid G_2 phase advances the subsequent mitosis.

The Organism: Physarum polycephalum - We have chosen to work with the lower
eukaryote myxomycete, Physarum polycephalum, having the following life stages:
A haploid spore stage which germinates to a haploid amoeba capable of mitotic
division. The amoeba can encyst as a resting haploid state. In a liquid
medium, the amoeba develops a flagellum, which it loses on return to solid
medium. Many mating types exist. Two amoeba of appropriate mating types
can fuse in syngamy to form a diploid zygote. Homothallic mutants exist
allowing an amoeba to mate with a sister. The zygote grows by repeated
nuclear division, without intervening cell division, to form a macroplas-
modium containing up to 10^{10} nuclei in a common cytoplasm. In adverse circumstances, the plasmodium follows two pathways of differentiation; to a
diploid resting sclerotial stage, or to the haploid spore stage.

The plasmodium of Physarum is among the best possible preparations for
study of the mitotic cycle. Well plated plasmodia exhibit virtually complete
spontaneous mitotic synchrony, in which up to 10^{10} nuclei go through metaphase
within two minutes of one another during a 10 hour cell cycle. The surface
plasmodium grows up to 10 to 15 centimeters in diameter. Massive nuclear

cytoplasmic transplant experiments can be performed by slicing a crescent off
of two plasmodia, and abutting the cut crescents. Fusion of the plasmodial
membranes occurs within about 40 minutes, then exchange of cytoplasm between
the two pieces is brought about by extensive cytoplasmic streaming generated
in an anastomotic net of "veins" which form across the join. By labeling
nuclei of one fusion partner, we have shown that on the order of 30% of the
nuclei cross from one plasmodia piece to the other fusion partner in about
2 hours. Such plasmodial fusion allows massive mixing of nuclei and cytoplasm
from any two well defined stages of the mitotic cycle. The mitotic sequence
lasts about 30 minutes, and can be conveniently followed in squash preparations
viewed under oil in the phase microscope.

The Oscillation of the Mitotic Clock: Limit Cycle Model

To explain the predictions we have made and tested (47,48) requires
describing several properties of continuous biochemical oscillations. For
simplicity, we consider only our own hypothetical biochemical oscillation.
We suppose that during the Physarum mitotic cycle, a division protein, X, is
synthesized at a constant rate A. X converts to an activated form, Y, without
further protein synthesis, at a rate proportional to X, BX. Y catalyzes its
own formation from X at a rate proportional to X and Y^2. For example, catalytic
conversion of X to Y might require activation of an enzyme by binding 2 Y
molecules. Y decays at a rate proportional to its concentration. Finally
we assume a threshold level of Y,Yc triggers mitosis.

There are no biochemical grounds to believe this particular kinetic
scheme is correct, and we wish to stress that at this stage of our knowledge,
the importance of our model is that it is a generic example of a kinetic
system capable of biochemical oscillations, which has the same critical
properties as more realistic examples. We pick this particular kinetic
scheme first because the feedback activation of Y on its own production parallel
the feedback activation by the product of phosphofructokinase on the enzyme,

generating the glycolytic oscillation (34). Second, it is the simplest
kinetic scheme giving rise to sustained oscillations. Furthermore, it generates
a wave form which characterizes the control system underlying Physarum mitotic
control. In formulation, it is rather similar to the simple division protein
model which has dominated the field for so long. This model was formulated
prior to Bradbury's et. al. (43,44) report on the F_1 histone phosphorylation
role in controlling mitosis, and their contention that this biochemical is a
good candidate for a variable of the mitotic clock. With slight alterations,
we could interpret F_1 histone as the X of our model, and Y, the activated
product of X, as the phosphorylated state of F_1 histone. As we are about to
show, our kinetic scheme gives rise to a sustained oscillation of X and Y
concentration. A similar sustained oscillation of the concentration of F_1
histone, and phosphorylated F_1 histone would constitute one possible more
realistic kinetic scheme having the same critical general properties.

Our kinetic scheme defines a pair of differential equations linking the
synthesis and degradation of X and Y.   (see appendix).

1. $dX/dt = A - BX - XY^2$

2. $dY/dt = BX + XY^2 - Y$

The wave form of X as a function of time, and Y as a function of time
are shown in Figure 1. The gradual increase of X, and its rather rapid drop
near the time of mitosis are similar to the sawtooth wave form commonly
assumed for the mitogen model. In Figure 2, we plot the concentration of Y
at each moment of time, against the concentration of X at that moment. This
plot gives the state of the biochemical system, that is, the concentrations
of X and Y, at each instant of time. As the concentrations change, the point
representing the simultaneous concentrations in the XY "state space" changes
continuously. If the system is undergoing a sustained oscillation, it must
come back to the same state again after one period, which will yield a closed
path such as the closed, roughly triangular path in Figure 2. As time goes

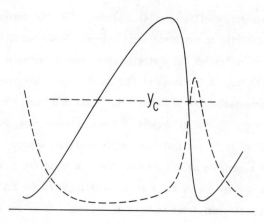

1. Time wave form of X and Y concentrations generated by our model mitotic
    oscillator for A = .5, B = .05.

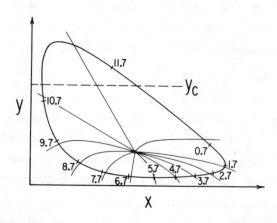

2. The limit cycle in the concentrations of X and Y. A critical level of Y,
    Yc, triggers mitosis. The nearly radial lines emanating from a point inside
    the cycle are "isochrons" separately equal intervals of time along trajectories,
    and give hours before mitosis along the limit cycle. All isochrons meet at
    the steady state singularity, S, inside the limit cycle.

forward, the point representing the system travels counterclockwise around
that path. A horizontal line, Yc, shows the level of Y needed to trigger
mitosis. A central feature of our model is that mitosis is _not_ part of the
mitotic clock. Were mitosis suppressed, the XY oscillation would continue.
Along the cyclic path, we have marked off points at successive twelfths of
the total period, representing hours before mitosis.

If the model oscillation system is released from any specific concentrations of X and Y, or _state_ in the XY state space, the system follows a
unique trajectory, or path, through its state space. Since the system is
deterministic, no trajectories can cross. However, some state might not
change at all with time, but might instead be a steady state. A fundamental
property of all real, continuous biochemical oscillations (45) is that they
occur as rotations around at least 1 steady state, S. Setting equations 1
and 2 to 0, we find the steady state of our model to be $X = A$, $X = A/(B + A^2)$.
If the system is released at that combination of concentrations, it will
remain forever unaltered. This feature of all continuous oscillations is a
critical distinction between our model and either the discrete loop of states
model, or the familiar division protein model. Neither of these occur as rot-
ation around a steady state. Many predictive consequences follow from this
difference. As noted above, we divided the cyclic path into 12 hours before
mitosis. Hence _phase_ of the cycle is well defined on that cyclic path. But
at the steady state, S, no oscillation is occurring. Therefore, S has no phase
at all. Unlike the closed loop sequence of states model, or the division
protein model, the biochemical oscillation model has a set of states with
well defined phases, surrounding another state with no phase.

In our model, if a system is released from a state near S, it follows a
trajectory which spirals out to the closed cyclic path; if released from an
initial state outside the closed path the system spirals onto the closed
cyclic path, reaching the closed path as time goes to infinity. Such a closed

path is called a limit cycle. Its critical feature is that from almost any
initial condition, the system ends up on the same cyclic path, and therefore
exhibits the same wave form and period biochemical oscillation. This type of
stability to perturbations is a natural prerequisite for any model of a
stable periodic phenomenon like mitosis in Physarum.

Figure 2 also shows 12 nearly radical lines emanating outward from S,
and crossing the limit cycle. These lines, or isochrons (46) separate equal
1 hour intervals of time along trajectories. Where isochrons are closely
spaced along the limit cycle, angular velocity is slow, where isochrons are
far apart, angular velocity is high. An important property of isochron is
that if a number of identical copies of the oscillating system were simultaneously released from many points on one isochron, they would all eventually
synchronize into phase with one another as they spiral onto the limit cycle,
while they continue to rotate. Thus each isochron gives loci equal ultimate
phase. All isochrons terminate on the phaseless steady state S. Thus two
states near one another but on opposite sides of S, spiral out to opposite
sides of the limit cycle, far out of phase with one another despite being
near one another and S initially. Therefore, we can see that a small volume
of states surrounding the steady state S, will spiral out to all phases on the
limit cycle and therefore the small volume has representatives of all ultimate
phases. An equal small volume of states centered on a point on the limit
cycle, represents only a few ultimate phases. Thus, if identical copies of
our biochemical system were released from the small volume of states surrounding
S, they would all wind out to the limit cycle completely out of synchrony with
one another. But if copies of the system were released from a small set of
states near the limit cycle, they would wind onto the cycle in near synchrony.
This is a critical distinction between our, or any, oscillation model which
rotates around a set of states near S having no, or equivalently all, phases
and the classical closed loop sequence of states model, or the division

otein model, both of which are topologically equivalent to a one dimensional
osed ring of states. In the ring case, phase is identical to state on the
ng, and the ring sequence does not surround a phaseless state. For the bio-
emical oscillation phase is not identical to state on the limit cycle. All
tates on the same isochron have the same phase. Figure 3 shows the same model,
it with the numbers of hours from each state until the first crossing of Yc,
nd hence the first triggering of mitosis. A state near the steady state S,
ollows a trajectory which spirals out to the limit cycle, taking several
otations to gain sufficient amplitude to cross Yc and trigger mitosis.
herefore the state space is divisible into concentric rings of states

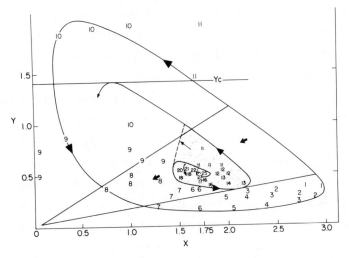

3. Number of hours various states of the model oscillator are from the next
mitosis. Those states which are greater than one cycle from mitosis lie
in a concentric zone surrounding the steady state (shaded zone). The set of
possible states inside the limit cycle which can be reached with increasing
durations of heat shock applied from a locus on the limit cycle about 40
minutes prior to mitosis are shown by the path of the solid arrows. Yc =
critical threshold level of Y needed to trigger mitosis.

centered on the steady state, successively 0 to 1 cycle times from crossing
Yc, 1 to 2 cycles from crossing Yc, 2 to 3 cycles from crossing Yc. Very
near S, the system is indefinitely many cycles from crossing Yc.

Predictions Already Tested and Confirmed

A large set of experiments to test this model were performed using heat
shocks at 38°C. The normal growth conditions is 23°C. We assume that the
dominant effect of a 38°C heat shock is to destroy X, and Y, at a moderate
rate, proportional to their concentrations. Then a heat shock drives the
system off the limit cycle, roughly toward the origin, $X = Y = 0$. The effect
of destroying 20% of X and Y at each phase of the limit cycle oscillation is
shown in Figure 4A. The model makes a number of striking predictions.

1. Soon after mitosis, a heat shock drives the oscillation more rapidly
in the direction it was heading, which should result in a phase advance of
the subsequent mitosis.

2. After about 3 to 4 hours post mitosis, a heat shock drives the system
backward toward the origin, and should result in a delay. As G_2 advances the
delay induced by the same heat shock should increase since the destruction
"vector" is longer, representing a large retrograde movement, and because
isochrons are packed closer together in the last few hours before mitosis.

3. In the last hour before mitosis, the system has "rounded the corner"
and is moving up the hypotenuse of the triangular limit cycle path. A short
heat shock should displace the system inside the limit cycle a short distance
but leave it on the same isochrons, from which it should follow a trajectory
to cross Yc on schedule. Hence for short heat shocks, the delay in mitosis
should increase from about 4 hours after mitosis, reach a peak about 1 hour
before mitosis, then should decline to almost nothing.

4. A very striking prediction is that if a longer heat shock drives
the system further toward the origin, $X = Y = 0$ than does a short one, then a
longer shock just before mitosis will drive the system inside the limit cycle

into a ring of states 1 to 2 cycles from mitosis. Hence, as heat shock duration increases in very late G_2, mitotic delay should suddenly jump from nearly no delay, to a bit more than a full cycle delay. In contrast, as heat shock duration increases for shocks applied at another phase, about 4 hours after mitosis, the system is driven outside the limit cycle and toward the origin, rather than inside the limit cycle. Thus, no such discontinuous jump in delays should occur.

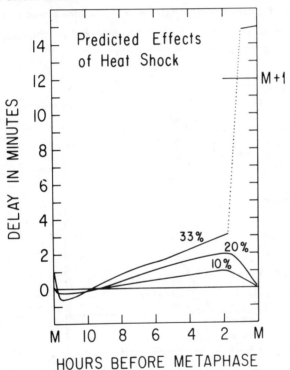

4A.) Predicted effects of destroying 10%, 20%, and 33% of X and Y as a function of phase in the cycle. Shortly before mitosis a 10% destruction yields a slight delay, a 33% destruction yields a full cycle delay. Shortly after mitosis destruction of 33% of X and Y causes a phase advance. Ordinate: delay in hours.

Figure 4A shows these predictions for 10%, 20% and 33% destruction of X
and Y for all phases of the oscillation. In Figure 4B, we show the result
of a large number of heat shock experiments at all phases, in which the heat
shock duration ranged from 1 to 180 minutes. We show the delay from the <u>end</u>
of the heat shock, to mitosis, compared to an unshocked piece of the same
plasmodium. Each point is a distinct experiment. All predictions are
confirmed. 1 and 10 minute shocks shortly after mitosis give the predicted
phase advance in the next mitosis. Mitotic delay reaching a peak in late
G_2, and declining to no delay for 1 and 10 minute shocks was observed. For

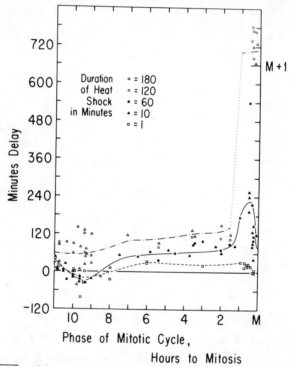

4B.) Confirmation of predictions with 1 min. to 180 min. heat shocks at 38°C
to Physarum. A discontinuous jump to a full cycle delay occurs in late G_2 as
heat shock duration increases; phase advances occur shortly after mitosis.

1, 2, and 3 hour shocks, the delay gradually increases as G_2 advances, but discontinuously jumps to a bit more than a full cycle delay as predicted in very late G_2. One of the few failures of our model is the delay induced by long heat shocks shortly after mitosis, where we predict in advance. We believe that long shocks at this phase disrupt S, thereby inducing the unexpected delay.

Several points are worth stressing. We have confirmed earlier results by Brewer and Rusch (28) showing that in Physarum, heat shocks give no delay early in the cycle, and give a delay increasing to a peak in late G_2, then declining as mitosis approaches. This form of delay curve is very widespread, having been found in Tetrahymena (3,4) Physarum, and even mammalian cells (9). It has been a cornerstone for the division protein model of Zeuthen, in which the gradual accumulation of a heat labile division protein, and the late G_2 decline in delay is taken to reflect the rapid conversion to a heat stable structure. The absence of delay in the early part of the cycle is uniformly interpreted to mean no "mitogen" is synthesized at that phase. We have predicted exactly the same form of "variable excess delay curve" directly from the topological properties of a simple biochemical oscillation. This, at least, casts serious doubt on the familiar interpretation. Further, the confirmed phase advance induced by early shocks are difficult to interpret in the familiar division protein model. Also, that model is hard pressed to explain why, for short shocks as G_2 advances, delays peak then decline, while for long shocks, delay suddenly jumps to a full cycle delay. The division protein model would have to assume that the heat stable division structure is stable to short shocks, but destroyed all or none by long enough shocks. While possible, this is an ad hoc addition to that theory, but a predictive consequence of our oscillation model. Our model has made a number of additional predictions which have been confirmed.

5. After a late G_2 shock, the system is driven inside the limit cycle, and its crossing of Yc is delayed, hence mitosis is delayed. But the system crosses

Yc at a point <u>inside</u> the limit cycle. From that point, the distance around to the next crossing of Yc is <u>less</u> than the normal- full distance around the limit cycle, hence the system should catch up on the second cycle post shock. We and others (28) have confirmed this prediction.

6. The most striking set of predictions can be seen from Figure 3, which shows the concentric rings of states centered on S, successively 0 to 1, 1 to 2, 2 to 3, etc., cycles from crossing Yc and undergoing mitosis. If longer heat shocks drive the system closer to the origin, X=Y=0, then in very late G_2 shocks should drive the system inside the limit cycle, and as heat shock duration incre the biochemical system should be driven further and further into the limit cycle, crossing into zones part of a cycle from mitosis, one cycle from mitosis, two cycles from mitosis, perhaps three cycles from mitosis. Then for even <u>longer</u> shocks, the system should be driven <u>past</u> the steady state, S, toward the origin, back out through the rings of states until it emerges in the outer ring nearest the limit cycle and the origin, where the mitotic delay has been reduced from 2 or more full cycles, to only about nine hours. As heat shock duration increases for very late G_2 shocks, delay should first increase up to about 2 or 3 cycles, then dramatically decrease to less tha a cycle. By contrast, the same shocks applied four hours after mitosis should result in a monotonically increasing delay, which should never be more than a full cycle, for the system is driven <u>outside</u> the limit cycle, toward the origin.

In figure 5A we show the results of many heat shocks applied for 3,6,9,or 12 hours either within 40 minutes of mitosis (late G_2), or four hours after mitosis. All these predictions have been confirmed. Consider first the late G_2 shocks. Each point represents only the first synchronous mitosis seen in one plasmodium after its heat shock. Three hour shocks resulted in delays of about 4-5 hours, 11 to 12 hours, or 22-23 hours. Thus delays of part of a cycle one cycle, or two full cycles are seen, for the 11 and 22 hour peaks represent successive cycle periods. In some of these plasmodia, no evidence of mitosis was seen for two full cycles, then synchronous mitosis observed. Thus, it appears that Physarum can skip mitosis for two full periods, yet keep count of cycle times up to two. This is one powerful line of evidence that mitosis can be suppressed in Physarum, but its clock can continue to keep time, and therefore that mitosis is not a necessary part of its own clock. Further evidence supporting this will be described below.

For late G_2 shocks, as heat shock duration increases to 6 and 9 hours, delay to mitosis increases up to 2 cycles, but for 12 hours shocks, the delay decreases dramatically to about 9 hours, as predicted. Thus for late G_2 shocks, the delay has increased, then decreased as shock duration has increased. By contrast, shocks, 4 hours after mitosis, show no such inversion as heat shock duration increases, Fig. 5B. Delays increase from 1 to 7 hours as shocks increase to 12 hours. Plasmodia do not go through mitosis in 38°C.

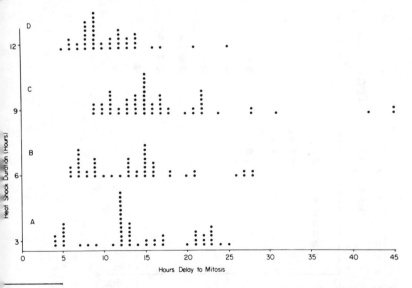

5A.) Results of 3,6,9 and 12 hour heat shocks applied within 40 minutes of mitosis.

We feel this is a striking confirmation of our general theory. It cannot be
predicted at all by the loop sequence of states model, or familiar division
protein model. Some conceivable ad hoc postulates might be appended to each
however. For example, one could say that temperature accommodation occurred
after 6 hours in late G_2, but never in early G_2, so in late G_2 plasmodia heated
longer than six hours, "phase" goes slowly forward, resulting in less delay.

We tested our oscillation model in an entirely different way, by plasmodial
fusion at all possible phases and phase differences. We model plasmodial
fusion as the fusion of two boxes, each containing a copy of our X Y biochemical

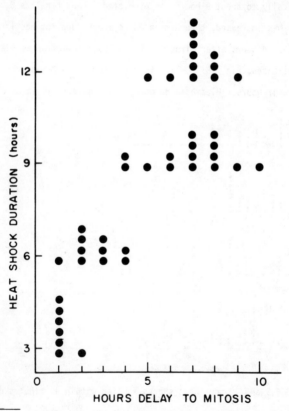

5B.) Effect of 3,6,9 and 12 hour heat shocks applied 4 hours after mitosis.

cillation, and separated by a semipermeable membrane which allows X and Y to

ffuse between the boxes at a rate proportional to the concentration differences.

Fig. 6A and 6B,we show the results of starting the "diffusive mixing",

ken to model pumping through the anastometic net of veins across the plasmod-

l fusion join, for fusions at two different phases and phase differences. In

e first, the trajectory of the retarded plasmodium, A, crosses Yc roughly on

nedule. Its mitosis is about on time. In 6B, the initial phase difference

larger, and A's trajectory is "tugged" inside the limit cycle by B ahead of

, such that A curves below Yc, misses mitosis, and is subject to a long delay

it must cycle around again. A critical prediction, therefore, is that if

e B member of the fused pair is shortly before mitosis, and A is chosen

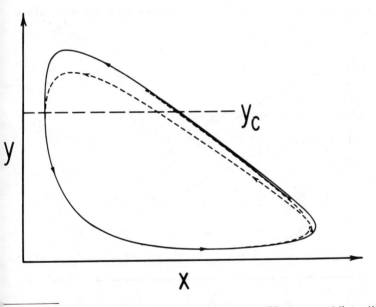

A.)"Fusion" of two mitotic oscillators, A and B by allowing X and Y to diffuse

etween A and B, when B is 0.7 hours prior to mitosis and A is 1.7 hours prior

o mitosis. Both the trajectory from A and from B cross Yc roughly on

chedule. Both A and B go through mitosis roughly on schedule.

successively earlier in the cycle for fusion with such a B, then A would at first suffer little delay while its trajectory crosses Yc on schedule, then A should discontinuously jump to show a very long delay when it is so early with respect to B that its trajectory curves below Yc. In Figure 7A we show the results of many such fusions at all possible phases and phase differences. A (-) indicates no significant delay in A's mitosis. A (+) indicates a very long delay in A's mitosis, from 2 to 8 hours. For comparison, in Figure 7B, we record the predictions of our model. In both cases, the locations of the delays, +, are identical, and both show the discontinuous jump from no delay to long delays as the phase difference increases from less than 80 minutes to more than 100 minutes, for fusion with a B taken near its time of mitosis.

The familiar sawtooth oscillations does not even yield synchronization.

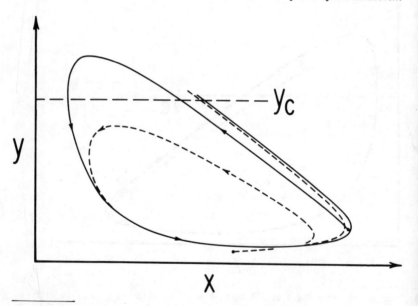

6B.) "Fusion" of B at 1.7 and A at 5.7 hours prior to mitosis. B crosses Yc but A curves below Yc and misses mitosis. Nevertheless, the mitotic oscillator continues to cycle.

Further, that model predicts that as phase difference increases, mitotic delay increases monotonically for the retarded member, while we found no delay and a sudden jump to a long delay as the phase difference increases. The closed loop sequence of states model has no coherent predictions to make concerning these phenomena.

In our fusion experiments, we discovered a new phenomenon, which we call abortive prophase. Here nuclei enter the early stages of prophase, but the nucleoli do not totally disperse as they normally do, rather they break into several small clumps. No metaphase is seen, yet the nuclei reform a larger

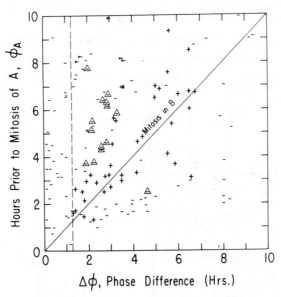

7A.) Results of plasmodial fusions on the mitosis of plasmodium A. Time before A's mitosis is plotted vertically, phase differences between A and B, horizontally "+" means a delay in A's mitosis of two or more hours; "-" means a delay less than two hours; "Δ" means abortive prophase; "o-","-o" means a dominance by one plasmodium (A and B respectively) of the phase to which the AB mix synchronizes.

than normal interphase nucleus and appear to enter S, at least by the criteria
of H^3 Thymidine autoradiography. We scored these abortive prephases with trian
in Figure 7A. These locations correspond, in Figure 7B , to fusion in which the
trajectory of A neither passed above Yc, nor clearly below it, but just grazed
Yc. We therefore suppose that abortive prephase is induced by not quite reachi
threshold in Y concentration during the oscillation. In terms of the interpretation of our XY oscillation as F_1 histone, and its phosphorylation, we would
say that an abortive prophase is the consequence of a slightly less than adequa
level of F_1 histone phosphorylation.

Measuring the Wave form of the Mitotic Oscillation

We have introduced a technique to characterize the waveform of the dynamic

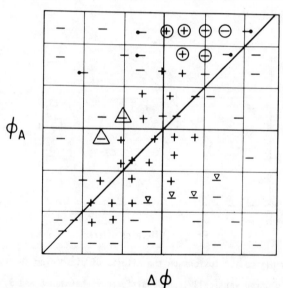

7B.) A phase-phase difference plot of mitotic blocks of one model mitotic
oscillator when coupled by diffusion to a second model oscillator. "0"=
was also blocked. The other symbols, "+", "-","Δ", "o-", and "-o", carry the
same meaning as in Figure 8.

system underlying the periodicity of mitosis in Physarum, which depends upon the fact that the synchronized phase of a fused pair of plasmodia must show certain discontinuities as the phases fused are varied systematically. Measuring Mitotic Oscillator: The Arc Discontinuity, (47,48) describes this method. The results for Physarum strongly suggest that the dynamical system comprising the mitotic clock has a waveform fundamentally similar to that of our simple two variable XY oscillation model.

Is Mitosis Part of the Mitotic Clock?

Two lines of evidence suggest that mitosis is not itself part of the clock timing its own occurrence in Physarum. 1. Three hour heat shocks applied in late G_2, resulted in some plasmodia which skipped mitosis for two full cycles with no signs of the mitotic sequence in any nuclei. The apparent rhythmicity of the three hour late G_2 shocks results suggests the plasmodia are keeping time up to two periods with no intervening mitosis. This implies that mitosis is not a necessary part of the capacity to account two periods. It is a predicted consequence of our model, in which mitosis is downstream of the oscillation clock. 2. In the same 3 hour late G_2 shocks, in some plasmodia we saw a partially synchronous mitosis 7 hours before the fully synchronous mitosis which we scored in the figure. The fraction of nuclei participating in the synchronous mitosis ranged from 5% to 90%, but had virtually no effect on the seven hour duration until the following fully synchronous mitosis. If it were the case that passing through mitosis set the nucleus, or cytoplasm in its vicinity to some 0 state, one would predict a longer delay to the fully synchronous mitosis following a partial mitosis in which 90% of the nuclei participated, than one in which 5% participated. This would be the direct prediction of the sawtooth mitogen model, but it was not observed. This suggests that the clock is independent of mitosis itself.

Preliminary evidence suggests that DNA synthesis is also not part of the mitotic clock. In several heat shocked plasmodia which skipped one full cycle

we found no evidence of DNA replication by H^3 Thymidine autoradiography.
We have not yet attempted to show Physarum can count up to two cycles without
an intervening S period.

Evidence of Contact Inhibition of Mitosis by Plasmodial Fusion

Our detailed model, and the general concept that the mitotic clock is
partially characterized as a continuous dynamical oscillation in which a
threshold level of one or more variables triggers mitosis, strongly suggests
the possibility of transient or persistent subthreshold oscillations induced
by diffusive, or other, coupling between the oscillation variables in adjacent
cells. We model plasmodial fusion by juxaposition of two identical boxes
containing identical copies of the XY biochemical oscillation system, and allow
X and Y to diffuse between copies according to Fick's law. We find that fusion
at certain phases tends to cause both fused boxes to oscillate below the Yc
threshold for two or three cycles, each box "tugging" the other inside the
limit cycle, and below Yc. After a few cycles, the systems synchronize
and wind out to the limit cycle. This behavior occurs when the diffusion
constants for X and Y are nearly equal. We found a startling behavior when
the diffusion constants are strongly unequal, and Y is nearly non-diffusible,
while X diffuses well. (see appendix) If boxes having similar phases were
"fused", the model pair of cells synchronized rapidly. If two boxes far out of
phase were fused, then the pair nearly stopped oscillating and the pattern of
oscillation in box 1 differs from that in box two. An intuitive
understanding of this phenomenon is that when the diffusion constants of X and
Y are nearly identical, then at the time of fusion with box 1 in state X_1 Y_1,
and box 2 in state X_2 Y_2, the diffusive force will tend to pull each point
representing one box in the XY state space, directly toward the other point
along the line connecting them in XY state space. Hence, they tend strongly
to synchronize. But suppose only X can diffuse, not Y. Then upon fusion
the two points representing the two boxes in XY state space are pulled by

iffusive forces <u>parallel</u> to one another, with no change in the Y dimension, s they move toward one another in the X dimension. The two points tend not o come together in XY space, and thus tend not to synchronize. The consequence f that failure will tend to be the disruption of the spatially homogeneous imit cycle oscillation, and the appearance of some new pattern of stable ynamical behavior, such as collapse to a steady state, or a new inhomogeneous attern of biochemical oscillation. If so, crossing of a threshold like Yc is ikely to fail, and an enduring inhibition of mitosis by contact would result.

Thus, from the simple hypothesis that the mitotic clock is partially a ontinuous oscillation with a threshold level of some variable triggering itosis, we deduce that fusion allowing diffusive coupling (or coupling by a ariety of other "transponding" signals between non-fusing adjacent mammalian ells) can lead to transient suppression of mitosis. And if some variables f the oscillation diffuse much more easily than others, fusion is likely to ause cessation of significant amplitude oscillations hence induce stable ontact inhibition of mitosis. A critical consequence is that merely by ffecting a <u>gradual alteration</u> in the ratio of diffusive (or other coupling) oefficients of oscillation constituents, a cell system can suddenly switch rom one which tends to synchronize and allow continued mitosis, to one which nduces contact inhibition of mitosis. We think this is a very important new, imple, and plausible general mechanism to account for contact inhibition of rowth, regardless of the particular molecules which happen to be the constit- ents of the cellular oscillation.

We already have preliminary evidence that fusions at appropriate phases an suppress mitosis for several cycles. In some fusions inducing abortive rophase, normal mitosis was suppressed in both partners for up to four full ycles.

ssessment of the Model

It seems fair to say that our model has had several predictive successes.

It correctly predicts that heat shocks early in the cycle produce a phase advance, that for short shocks, as G_2 advances, mitotic delay increases to a peak and declines to nothing; that long shocks in late G_2, but not early G_2, cause skipping of one or even two mitoses while its clock appears to tick; that for late G_2 shocks, as the shock duration increases, mitotic delay first increases, then decreases dramatically; it correctly predicts the results of very many plasmodial fusion experiments; and preliminary evidence supports the prediction that fusion of appropriate phases can suppress mitosis in both partners for several cycles. In its simple form, the closed causal loop of discrete states model of the cell cycle simply cannot make any of these predictions. In its simplest form, the single variable mitogen model, whose threshold level of mitogen triggers mitosis and is caused by mitosis to fall to a low level, is unable to account for these results, without having been a to predict them.

We would stress that we view our specific mathematical model merely as a generic example of an oscillating biochemical system with roughly similar properties such as wave form, period, and stability to perturbations. There are no compelling grounds to believe that only two variables like our X and Y say F_1 histone and phosphorylated F_1 histone, are involved in timing mitosis, or that no discrete events of the mitotic cycle are parts of the oscillation itself. A more adequate general view would picture a complex web of biochemi conversions, some behaving as continuous variables, some as discrete processe which switch on and off, all embedded in an overall oscillation which is tuned by many internal parameters. This is a more creditable general view t either the simple causal loop model, or the single substance, sawtooth, mitog model.

The importance of introducing the techniques of non-linear dynamical systems is that they provide a coherent conceptual framework in which to describe the integrated dynamical behavior of such systems, and allows us to devise experiments to characterize those integrated behaviors. Non-linear dynamical systems typically partition their state spaces into subvolumes called basins of attraction. If the system is released inside any basin of attraction, it remains there, and flows to some asymptotic "attractor set"; for example, all states in the basin may flow to a single stable steady state, or to a limit cycle. Each basin of attraction has certain other basins as nearest neighbors in its state space. The importance of this is that one way to change the behavior of a dynamical system is to perturb it from one basin into a neighboring basin, hence it might switch from an oscillation, in the first, to a stable steady state in the second - say from the mitotic cycle to the non-dividing G_0-state. We note that even our very simple model has two basins of attraction: the single steady state S, is the first, and all the rest of the state space which flows onto the limit cycle is the second. One possible location, in our oscillation theory, for the non-dividing G_0 state, would be at the non-oscillating steady state S. A very important feature of even our simple model is that as parameters of the XY system gradually change first the waveform, period, and amplitudes of oscillations are deformed, then abrupt alterations in the topology of the dynamical patterns of behavior of the XY system occur. Oscillations appear and disappear, new steady states appear and disappear. Parametric tuning is a second critical way cells might control their dynamical behavior. For example, in our simple model, as parameter A is tuned to lower values, the amplitude of the limit cycle <u>contracts</u>. At a critical value of A, the oscillation has contracted to a steady state point which has now become stable. Hence, simply by tuning a parameter analogous to A, the cell could control entry into a non-dividing G_0 state.

Without the concepts of nonlinear dynamical systems, this hypothesis could hardly be stated. It is our belief, that these integrated dynamical patterns of behavior of the mitotic clock, and their alterations as parameters of the system are varied, constitute the appropriate macroscopic biological

observables for the cell system, for these _are_ the integrated behaviors of
the cell.

REFERENCES

1. Wu, A.C., and Pardee, A.B., J. Bact. 114 (1973) 603.

2. Smith, H.S., and Pardee, A.B., J. Bact. 101 (1970) 901.

3. Zeuthen, E., and Rasmussen, L., Research in Protozoology (ed. T.T. Chen)
   Vol. 4, P.9 Pergamon Press, New York (1972).

4. Zeuthen, E., Synchrony in Cell Division and Growths (ed. E. Zeuthen)
   Interscience, New York.

5. Resmussen, L., Exptl. Cell Res. 48 (1967) 132.

6. Kramhoft, B., and Zeuthen, E., C.R. Lab. Carlsberg 38 (1971) 351.

7. Rusch, H.P., Sachsenmaier, W., Behrens, K., and Grutter, W., J.Cell
   Biol. 31 (1966) 204.

8. Sachsenmaier, W., Remy, U., and Plattner-Schobel, R., Expt. Cell Res. 73
   (1973) 41.

9. Zeuthen, E., J. Cell Sci. 13, (1973) 339.

10. Mitchison, J.M., The Biology of the Cell Cycle, Cambridge Univ. Press,
    Cambridge (1971).

11. Hartwell, L., Science 183 (1974) 46.

12. Thormar, H., C.R. Trav. Lab., Carlsberg 31 (1959) 207.

13. Sachsenmaier, W., and Ives, D.H. Biochim. Z. 343 (1965) 399-406.

14. Rusch, H.P., Pred. Proc. Redn. Am. Socs. Exp. Biol. 28, (1969)
    1761-70.

15. Braun, R., and Behrens, K. Biochim. Biophys. Acta 195, (1969) 87-98.

16. Sylvan, B., Tobias, C.A., Malmgren, I.T., Ottoson, R., and Thorell, B., Expt. Cell Res. 16 (1959) 75-87.

17. Tauro, P., and Halvorson, H.O., J. Bact. 92 (1966) 652-61.

18. Tauro, P., Halvorson, H.L., and Epstein, R.L., Proc. Nat'l. Acad. Sci. 59 (1968) 277-84.

19. Gorman,J., Tauro, P., La Berg, M. J., Halvorson, H. Biochem. Biophys. Res.

20. Cottrell, S.F., and Avers, C.J., Biochem. Biophys. Res. Comm. 38 (1970) 973-80.

21. Eckstein, H., Paduch, V., and Hilz, H., Biochem. Z. 344 (1966) 435-45.

22. Kuenzi, M.T., and Fiechter, A., Arch. Microbiol. 64 (1969) 396-407.

23. Littlefield, J.W., Biochim. Biophys. Acta 114 (1966) 398-403.

24. Mittermayer, C., Rosselmann, R., and Bremesskor, J., Europ. J. Biochem. 4 (1968) 487-9.

25. Adams, R.L.P., Expt. Cell Res. 56 (1969) 55-58.

26. Gold, M., and Helleiner, C.W., Biochim. Biophys. Acta 80 (1964) 193-203.

27. Turner, M.K., Abram, R., and Lieberman, I., J. Biol. Chem. 243 (1968) 3725-8.

28. Brewer, E.N., and Rusch, H.P., Expt. Cell Res. 49 (1965) 79-86.

29. Doida, Y., and Okada, S., Nature, Lond. 216 (1967) 272-3.

30. Muldoon, J., Evans, T.E., Nygaard, O., and Evans, H., Biochim. Biophys. Acta 247 (1971) 310-21.

31. Howard, A., and Dewey, P.L., The Cell Nucleus (ed. J.S. Mitchell P. 156, Betterworth, London (1960).

32. Harris, H., Cell Fusion. Oxford Univ. Press, Oxford (1970).

33. Pavlidis, Theo. Biological Oscillators: Their Mathematical Analysis
    Academic Press, New York 1973.

34. Ghosh, A.K., Chance, B., Pye, E.K., Arch. Biochem. 145 (1971) 319.

35. Sel'Kov, E.E., Europ. J. Biochem. 4 (1968) 79.

36. Winfree, A.T., J. Math. Biol. 1 (1974) 73.

37. Goodwin, B. C. Nature (London) 209 (1966) 479.

38. Mano, Y., Biochem. Biophys. Res. Comm. 33 (1968) 877.

39. Mano, Y., Develop. Biol. 22 (1970) 433.

40. Oppenheim, A., and Katzir, N., Expt. Cell Res. 68 (1971) 224-6.

41. Sachsenmaier, W., Personal Communication.

42. Mohberg, J., Personal Communication.

43. Bradbury, E.M., Inglis, R.J., Matthews, H.R. Sarner, N., Eur. J.
    Biochem. 53 (1973) 131.

44. Bradbury, E.M., Inglis, R.J., Matthews, H.R., and Langan, T.A., Molecular
    Basis of the Contol of Mitotic Cell Division, submitted for publication,
    1974.

45. Glansdorf, P., and Prigogine, I., Structure Stability and Fluctuations
    Interscience Publ., Inc., New York (1971).

46. Winfree, A.T., Biological and Biochemical Oscillators (ed. B. Chance, E.K.
    Pye, A.K. Ghosh, B.Hess). Academic Press, New York (1973).

47. Kauffman, S.A. (1974) Boll. Math. Biol. 36, 171.

48. Kauffman, S.A. and Wille, J.J., J. Theor. Biol. (1975) 55, 47-93.

Lectures on Mathematics in the Life Sciences
Volume 9, 1977

SOME PROBLEMS OF FLUID MECHANICS IN BIOLOGY

S. I. Rubinow

In this paper we will present a brief review of work
performed in collaboration with Joseph B. Keller on some
fluid mechanical problems arising in biology. We shall try
to indicate areas of interest in which further research and
investigation would be most profitable.

It is worthy of note in this bicentennial year cele-
brating the start of the American Revolution to point out
that it also marks approximately the first mathematical
model of the flow of blood in the human circulatory system,
by, appropriately enough, the father of modern fluid dynamic
theory, L. Euler [1]. He considered an inviscid fluid flowing through a flexible tube of variable cross section with
an equation of state relating the pressure in the fluid and
the cross-sectional area. Euler complained of the difficulty of solving the equation system because of the non-
linear terms in it.

This single tube flow problem, which we can think of in
the first instance as representing blood flow in an artery,
has dominated mathematical considerations of the flow of
blood to this day. In recent times it was stimulated to a
notable degree by the work of Womersley [2]. It is usual

now to assume a more sophisticated and knowledgeable model of the mechanical behavior of the blood-artery system by treating the blood as a viscous, incompressible fluid and the arterial wall as (infinitely) long, thin, and visco-elastic. By assuming the blood flow and tube wall motion are of small amplitude, the equations of motion can be linearized. Then the mathematical problem is to solve the linearized Navier-Stokes equations for the fluid and the linearized equations of elasticity for the wall, with en-forcement of continuity of velocity and stress at the fluid-wall interface.

Numerous investigators, starting with Womersley, have proposed one or another variation of this model. If it is assumed that a longitudinal wave, harmonic in space and time, is propagating along the axis of the tube, the equa-tion system is readily solved. The entire motion of the blood and artery can then be analyzed into time harmonic waves of fixed frequency and wave length. The value of the propagation constant k is found as a root of the dispersion equation, in terms of the wave frequency ω and the other parameters of the problem. Because in this example the dispersion equation is very complex, involving Bessel func-tions of various arguments and kinds, its analysis contains the essential mathematical difficulties of the problem.

Practically all investigators have skirted the diffi-culties associated with the analysis of the dispersion relation by making the long wave length approximation. This limit does not account either for the full behavior of the waves, or for the difference in behavior between a wave

aving the fundamental pulse frequency and its harmonics.
sually only one mode of propagation has been examined, al-
hough there are infinitely many of them. These considera-
ions have motivated us to reconsider the single tube blood
low problem. In previous work [3], we considered the
roblem posed above for the case of an inviscid, compress-
ble fluid. The effect of the arterial imbedding in tissue
as represented by an impedance boundary condition at the
uter tube wall. The main effort of that work was the
tudy of the dispersion relation, by both analytical and
umerical means. The principal general characteristics of
he single tube system are as follows [4].

The dispersion equation has infinitely many roots or
ropagation constants, for a given harmonic frequency com-
onent. With each value of the propagation constant, there
s a set of profiles for the fluid pressure, fluid velocity,
nd tube wall displacement, called a mode. Each mode has
n amplitude, determined by the initial excitation of the
pplied pressure pulse. There are infinitely many propa-
ating modes, those for which k is essentially real, that
arry the energy of excitation. Two of these modes are
tube modes," and the rest are acoustic modes occurring at
ery high frequencies. We call one of the two tube modes,
hich are the ones of principal interest in blood flow,
oung's mode because the phase velocity at zero frequency
as first given by Thomas Young [5]. The other we call
amb's mode because its phase velocity was first investi-
ated by Lamb [6]. Associated with Young's mode (and the
coustic modes) is a resonance frequency for radial

oscillations of the tube, which we have suggested can help explain the phenomenon of post-stenotic dilation [4]. An arterial stenosis (narrowing) results in excitation of Young's mode near the radial resonance frequency, which tends to stretch the vessel and weaken it, leading to fatigue and a consequent ballooning out of the vessel downstream from the stenosis.

We have also considered the case when the fluid is viscous and incompressible, and the tube is free (zero outer impedance) [7]. The corrections to the behavior of the tube modes due to compressibility of blood are negligible. Figure 1 shows the behavior of the nondimensional phase velocity $c = d\omega/dk$ for Young's mode (denoted by subscript +), as a function of the nondimensional frequency ω, for various values of the nondimensional parameter α, proportional to the reciprocal of the coefficient of viscosity. Other tube parameters were chosen to be representative of mammalian arteries. Thus, the curve labeled ∞ displays the inviscid behavior of c_+. The ordinate value at zero frequency represents Young's velocity formula. Note that there is a "cut-off" frequency limit, beyond which c_+ is null, and the mode is no longer propagating. Of particular interest is the very low frequency behavior of c_+ that, at finite values of α, vanishes with ω. This behavior can be understood as a viscous boundary layer effect: The effects of viscosity are dominant at sufficiently small frequencies.

It is only for moderate, albeit small values of the frequency, in fact when $\alpha^{-1} \ll \omega \ll 1$, that c_+ is given approximately by Young's formula. For normal blood flow in the

human aorta, $\omega \sim 3 \times 10^{-3}$ (corresponding to the fundamental
frequency of the heart beat of about 70 cycles/min), and
$\alpha \sim 7 \times 10^{4}$, so that the above inequality criterion is readily
satisfied. We see in this set of circumstances the explana-
tion to the apparently paradoxical result that Young's
velocity formula, derived in the zero frequency limit with
blood considered as an inviscid fluid, agrees so well with
the observed values of the velocity of the pulse wave in
mammalian aortas, while the zero frequency velocity formula
for the case when the viscosity of blood is accounted for,
does not.

In figure 2 is shown the analogous behavior of the
phase velocity c_- of Lamb's mode. Here too, viscosity
alters the behavior of the mode at very low frequencies,
$\omega \ll \alpha^{-1}$. At higher frequencies, for which $\alpha^{-1} \ll \omega \ll 1$, c_- is
given in first approximation by Lamb's formula. Lamb's
mode, which has been largely ignored by blood flow investigators, possesses a profile showing tube displacements that
are relatively more important than pressure variations.
Because blood pressure rather than arterial motion is
usually measured, it is, with one exception [8], usually
not observed.

The arterio-venous part of the circulatory system can
be considered to be a network of elastic tubes containing a
viscous fluid, the behavior of which can be described in
the manner described above. A principal objective of the
study of such a model would be to determine the fluid
velocity, pressure, and tube motion throughout the network,
in response to prescribed pressure pulses or input

velocities at the end of one of the tubes of the network.
For example, the prescribed pulse could be that produced by
the heart at the entrance to the aorta. We have called this
model the wave-guide theory of the arterio-venous system, in
obvious analogy to its electromagnetic counterpart. The
theory can be simplified by averaging the pressure, velocity,
etc. over the cross section of a vessel, so that it becomes
a dual transmission line system: one line for Young's mode
and the other for Lamb's mode. The reason that only two
modes need to be considered is that the mammalian heart beat
is very small compared to the arterial wall resonance frequency. Even when several harmonics are considered, the
arterio-venous system is essentially a low frequency system.

It is possible to derive in a straightforward manner
[7] the following transmission line equations for the behavior of waves in any line:

$$-P_z = ZQ \quad ,$$
$$-Q_z = CP_t \quad , \qquad (1)$$

where P is mean pressure and Q is the mean volumetric flow
rate at a given position z in the tube at time t, Z is the
characteristic impedance and C is the characteristic com-
pliance of the tube. These latter two quantities are given
explicitly in terms of k, ω, and other parameters of the
tube. The partial derivatives with respect to z and t appearing in (1) are deceptive, because P and Q are understood
to depend on these variables as follows,

$$(P,Q) = (P_0,Q_0)\exp[i(kz-\omega t)] \quad , \qquad (2)$$

where the amplitudes P_0 and Q_0 are constants. If equation

(2) is substituted into (1) and the indicated differentia-
tions are performed, then we obtain the relation

$$-ik^2 = \omega CZ \quad , \tag{3}$$

which merely recapitulates the dispersion relation.

Hence the passive mechanical behavior of the arterio-
venous system is viewed in this manner in complete analogy
with the behavior of an electrical network. Before such a
full theory of the arterio-venous system can be exploited,
the problem of what happens when a wave impinges on a junction of three tubes must be solved. This problem is un-
doubtedly the outstanding theoretical one that is delaying
the exploitation of the theory. However, there are many
other complications, such as lack of knowledge of the com-
pliance and impedance associated with the larger arteries,
the randomized nature of much of the network, and tube
leakiness at the capillary level. It is to be hoped that
in the future it will be possible to account for various
physiological observations for which there are no adequate
theoretical explanations at the present time. It should be
possible to assist the assessment of the state of health of
the circulatory system, as manifested by pressure and flow
observations at various positions, with the aid of the full
theory.

Because of the linearized nature of the theory described above, only small amplitude excursions from equilibrium are correctly described. There are many physiological
phenomena associated with the flow through elastic tubes for
which the collapsibility of the tubes must be accounted for.

A prime example is the steady flow of blood through a vein. The mean flux of blood Q through the superior vena cava of a dog was measured [9] as a function of the pressure difference $p_1 - p_2$, where p_1 is the pressure in the jugular vein (upstream from the superior vena cava) and p_2 is the pressure at the peripheral end. With p_1 maintained fixed and p_2 decreasing, it is found that Q initially increases linearly with $p_1 - p_2$, in agreement with expectations based on Poiseuille's law. However, with further decrease of p_2, Q rapidly attains an asymptotic level above which there is no further increase. Similar results are obtained for the flow through a rubber tube, and for the flow through the entire vascular system of the dog and through isolated cats' lungs.

We have proposed a simple theory for the steady flow through an elastic collapsible tube [10]. It is based on the assumptions that Poiseuille's law holds locally, and that at any location the radius of the tube is determined by the transmural pressure difference. It is also assumed that the stress-strain curve of the tube material is known. By means of these assumptions it is easily shown that Q is a function of L, $p_1 - p_o$ and $p_2 - p_o$ alone, where L is the length of the tube and p_1, p_2, and p_o are the upstream, downstream, and external pressures of the tube, respectively. Furthermore, Q is inversely proportional to L, is proportional to $p_1 - p_2$ when the latter quantity is small, and approaches a limit when $p_1 - p_o$ remains fixed and $p_2 - p_o$ approaches $-\infty$. When a stress-strain curve is introduced that is representative of the behavior of human arteries, then the calculated flow curves have the qualitative behavior of those observed, as described above.

The physical principles underlying the above theory can be readily understood if we think of a collapsible tube being used as a drinking straw. Initial increase of suction leads to increase in fluid flux, but beyond a certain point, no increase in flux is obtained because the favorable effect of an increased pressure gradient is negated by a decrease in the cross sectional area at the outlet.

Actually, a decrease in pressure to a level sufficiently below the value of the external jacket pressure leads to buckling of the tube, which was not taken into account by the above theory. When buckling occurs, the cross section of the tube is no longer circular, and the flux through the tube is thereby decreased. If the shape of the cross section is known, the flux can be determined by solving a suitable flow problem. The shape of such collapsed tubes and the associated flow problem has been calculated [11], but the qualitative results of the theory are not significantly affected thereby.

The fact that an entire network of interconnected tubes is observed to behave as a single tube can perhaps be understood by the following considerations. Imagine a ladder network consisting of two large diameter flexible tubes connected by N identical small diameter flexible tubes in parallel. If the pressure drop along the large tubes is negligible, the pressure in each of them can be considered to be constant. Then the flux through the network is N times the flux through a single small tube.

There are other phenomena involving flow through flexible tubes which involve large amplitude displacements varying in time. For such phenomena, neither the steady state theory nor the linearized theory is adequate. We mention as examples the Korotkoff sounds heard by the physician in making blood pressure measurements. Here an understanding of the dynamical transition between the state of negligible flow in a collapsed artery and the state of onset of flow that occurs when the external jacket pressure is decreased is required. In this transition vibrations of the tube known as "flitter" have been observed [12]. A second example is the coughing associated with air flow through the bronchi and trachea, in which tube collapse plays a vital role.

Another fluid flow problem unique to biology is the swimming of a microorganism with a single flagellum, such as a sperm which propels itself through the surrounding fluid by sending periodic bending waves down the flagellum away from its head. Two types of flagellar wave motion have been observed, helical and planar, with two corresponding types of trajectories of the organism. The observations have been sufficiently detailed to permit quantitative comparison with theory.

The fluid motion resulting from the flagellar motion is very slow, having a Reynolds number of 10^{-3} or less. Hence the theory of slow flow is applicable to it, and was introduced by Taylor [13]. He and other early theorists have largely confined their analysis to the determination of the forward or lingitudinal velocity component of the organism.

given that the tail motion is prescribed. Subsequently,
Chwang and Wu [14] calculated the longitudinal components of
the angular as well as the linear velocity for the case of
helical flagellar motion. For arbitrary planar flagellar
motion, Brokaw [15] and others showed how to calculate both
components of the linear velocity and the single component
of the angular velocity.

We have reexamined the motion of such microorganisms
[16] because there are still some aspects of the observed
motions that have not been adequately explained previously.
We assume the flagellar motion is specified and an unknown
linear velocity $\vec{w}$ and angular velocity $\vec{\Omega}$ is assigned to the
entire organism. In terms of this prescribed motion, the
force and torque exerted by the surrounding fluid on the organism is calculated. From the condition that net force and
torque on the organism are zero, the velocities are found.
Knowing the velocities, the trajectory of the organism is
determined.

An immediate consequence of our analysis is that swimming motion requires relative motion of one part of an
organism with respect to another part. Thus, an organism
moving as a rigid body cannot swim. Hence, a helically
waving flagellum, which is a rigid body motion, produces
swimming motion because it is attached to a head that moves
relative to it. In fact, the head rotates in a direction
opposite to that of the tail. This explains why a headless
flagellum waving helically can not swim [14]. On the other
hand, a headless flagellum executing planar sinusoidal wave
motion, which is not a rigid body motion, can and does

propel itself [15]. In the case of planar, sinusoidal flagellar motion, we found that the trajectory of the organism
is a straight line with small oscillations about it. Each
point of the flagellum also oscillates longitudinally with
double the frequency of the transverse oscillations, producing a figure eight motion that has been observed. However,
if the planar oscillations are asymmetric, then the trajectory is a circle with small superposed oscillations, as observed in the tracks of sea urchin sperm and bull sperm.
The radius of curvature of the circle is inversely proportional to the magnitude of the asymmetry of the wave motion.

We derived a new formula for the instantaneous longitudinal component of the swimming velocity in this case, ex-
pressible in terms of elliptic functions of the first and
second kind. Illustrative calculations applied to sea
urchin sperm show that the velocity is slightly negative
during a short phase of its period. Based on it, the following approximate formula for the mean longitudinal velocity
component has been obtained,

$$-w_3 \approx c \; \frac{0.36k^2\rho^2}{1+0.72k^2\rho^2-(1+0.36k^2\rho^2)(3R/L) \; \log \; (b/\alpha L)} \quad . \quad (4)$$

Here w_3 is the longitudinal velocity component, positive in
the head to tail direction, c, k, and ρ are the phase velocity, wave number and amplitude respectively, of the wave,
b and L are the radius of the cross section and length,
respectively, of the flagellum, R is the radius of the head
(assumed to be a sphere), and α is a numerical factor. This
formula appears to be an improvement over that previously
given by Gray and Hancock [17]. We have compared equation (4)

with the observed swimming velocities of bull sperm executing planar sinusoidal motion [18] for which the values of
the parameters appearing on the right hand side of (4) were
measured or could be reasonably inferred. These observations, previously thought to be at odds with theory, were in
reasonable agreement with the predictions of equation (4).

In the case of sperm displaying helical wave flagellar
motions, our calculations of the trajectory show that the
organism moves along a helical path of small radius about a
straight line axis. Furthermore, the axis of the flagellum
is inclined at a small angle to the axis of the path, and
precesses about it. These conclusions are in qualitative
agreement with observations of long standing. The deriva-
tion of the longitudinal components of $\vec{w}$ and $\vec{\Omega}$ yields the
results of Chwang and Wu [14]. However, the agreement of
the theory with observations of bull sperm with helically
waving flagella [18] is not good.

One possible source of disagreement is that the motion
of bull sperm flagella is more complicated than that assumed
in the theory. Another possible contributing factor lies in
the calculation of the fluid dynamic forces on the tail
motion, which were based on the slender body theory ($b/L \ll 1$)
of Cox [19]. These forces are essentially those given by
Gray and Hancock [17]. In both calculations, the lift and
drag forces on an element of the slender body were assumed
to be local in character. The inclusion of higher order
terms in the theory of Cox may be necessary, to account for
nonlocal effects.

Many other slow flow phenomena occur in biology that

have received no theoretical attention or only recently have
begun to be considered by theorists. The following examples
constitute a partial listing: the swimming of ciliated
microorganisms [20], and of bacteria possessing many flagella
(in contrast to sperm, these flagella are known to be waved
passively from their base [21]); peristaltic pumping [22]
such as occurs in the flux of urine through the ureters,
elastic tubes connecting the kidneys to the bladder; axoplasmic flow, occurring in nerve axons, in which there is of
necessity a return flow in the same axon; other cellular
streaming phenomena and the puzzling observation of saltatory motion of cellular particles; amoeboid motion [23].
Applied mathematicians and fluid dynamicists seeking new
problems will find an examination of such phenomena reward-
ing. An extremely useful introduction into this entire area
of research is to be found in the recent work of Lighthill
[24].

GRADUATE SCHOOL OF MEDICAL SCIENCES
CORNELL UNIVERSITY
NEW YORK, N. Y. 10021

REFERENCES

[1] L. Euler (1775, published posthumously 1882).
"Principia pro motu sanguins per arterias determinando,"
Opera posthuma mathematica et physica anno 1844
detecta, ed., P. H. Fuss and N. Fuss. Petropoli: Apud
Eggers et socios, 2, 814-823.

[2] J. R. Womersley (1958). Wright Air Development Center,
Technical Report WADC-TR56-614 (compilation of all
publications of the author).

[3] S. I. Rubinow and J. B. Keller (1970). Wave propagation in a fluid-filled tube, J. Acoust. Soc. Amer. 50:
198-223.

[4] S. I. Rubinow and J. B. Keller (1968). Hydrodynamic
aspects of the circulatory system, in Hemorheology,
A. L. Copley, ed., Oxford: Pergamon Press, 149-155.

[5] T. Young (1808). Hydraulic investigations, subservient
to an intended Croonian lecture on the motion of the
blood, Philos. Trans. Roy. Soc., London, 98: 164-186.

[6] H. Lamb (1898). On the velocity of sound in a tube, as
effected by the elasticity of the walls. Manchester
Lit. and Phil. Soc. Memoirs and Proc. 42, No. 9.

[7] J. B. Keller and S. I. Rubinow, unpublished work.

[8] R. L. van Citters (1960). Longitudinal waves in the
walls of fluid-filled elastic tubes. Circ. Res. 8:
1145-1148.

[9] G. A. Brecher (1952). Mechanism of venous flow under
different degrees of aspiration. Am. J. Physiol. 169:
423-433.

[10] S. I. Rubinow and J. B. Keller (1972). Flow of a
viscous fluid through an elastic tube with applications
to blood flow. J. Theor. Biol. 35: 299-313.

[11] J. Flaherty, J. B. Keller, and S. I. Rubinow (1972).
Post buckling behavior of elastic tubes and rings with
opposite sides in contact. SIAM J. Appl. Math. 23:
446-455.

[12] J. P. Holt (1941). The collapse factor in the
measurement of venous pressure. Am. J. Physiol. 134:
292-299.

[13] G. I. Taylor (1951). Analysis of the swimming of
microscopic organisms. Proc. R. Soc. Lond. A Phys.
Sci. 209: 447-468.

138   S. I. RUBINOW

[14] A. T. Chwang and T. Y. Wu (1971). A note on the
     helical movement of micro-organisms. Proc. R. Soc.
     Lond. B. Biol. Sci. 178: 327-346.

[15] C. J. Brokaw (1970). Bending moments in free-swimming
     flagella. J. Exp. Biol. 53: 445-464.

[16] J. B. Keller and S. I. Rubinow (1976). Swimming of
     flagellated microorganisms. Biophys. J. 16: 151-170.

[17] J. Gray and G. J. Hancock (1955). The propulsion of
     sea-urchin spermatozoa. J. Exp. Biol. 32: 802-814.

[18] R. Rikmenspoel, G. van Herpen, and P. Eijkhout (1960).
     Cinematographic observations of the movement of bull
     sperm cells. Phys. Med. Biol. 5: 167-181.

[19] R. G. Cox (1970). The motion of long slender bodies
     in a viscous fluid. J. Fluid Mech. 44: 791-810.

[20] J. R. Blake (1971). A spherical envelope approach to
     ciliary propulsion. J. Fluid Mech. 46: 199-208.

[21] H. C. Berg (1975). Bacterial behavior. Nature,
     Lond. 254: 389-392.

[22] M. Y. Jaffrin and A. H. Shapiro (1971). Peristaltic
     pumping. Ann. Rev. Fluid Mech. 3: 13-36.

[23] G. M. Odell and H. L. Frisch (1975). A continuum
     theory of the mechanics of amoeboid pseudopodium
     extension. J. Theor. Biol. 50: 59-86.

[24] J. Lighthill (1975). Biofluiddynamics. Philadelphia:
     SIAM.

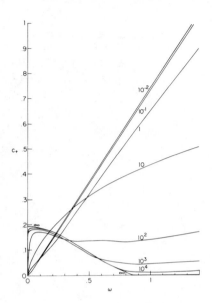

Figure 1. The phase velocity c_+ of Young's mode is shown as
a function of frequency ω, for various values of
the nondimensional parameter α, inversely
proportional to the fluid viscosity.

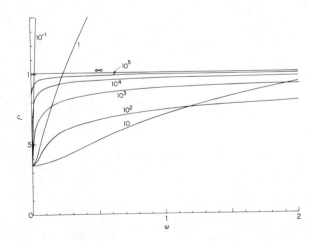

Figure 2. The phase velocity c_- of Lamb's mode as a function
of frequency ω, for various values of α.

Lectures on Mathematics in the Life Sciences
Volume 9, 1977

THEORIES OF AXOPLASMIC TRANSPORT

Garrett M. Odell*

ABSTRACT. The biological phenomenon of axoplasmic
transport and its significance are described. Following a
summary of characterizing data from the experimental literature and a survey of theories proposed to explain the
driving mechanism of axoplasmic transport, a new theory is
described in detail. It is hypothesized that waves of
conformational change propagate along the meshwork of submicroscopic protein filaments that pierce the cytoplasm
within nerve axons. The motion induced in the fluid which
bathes the protein meshwork is studied using a mathematical
continuum theory for generalized, moving, porous media. It
is shown that the right kind of filament motion can induce
a bi-directional flow up and down axons that could explain
many observed features of axoplasmic transport.

1. INTRODUCTION. In most living cells, the cytoplasm
moves. Sometimes this movement is associated with
macroscopic deformation of the entire cell -- as in the
case of macrophages, amoebae, fibroblasts, various cell
types migrating about within an embryo during morphogenesis
-- and sometimes cytoplasmic movement occurs within a cell
with more or less motionless boundaries. This article
comprises a description of a biological phenomenon,
axoplasmic transport, in which nerve cell cytoplasm
(axoplasm) moves within the stationary confines of a nerve
axon membrane (section 2), a brief survey of theories that

*Research supported by the National Science Foundation
under grant GP 39176 X.

have been advanced to explain the mechanism responsible for the motion (section 3), and a detailed exposition of my own ideas of how the driving machinery works (in the form of a mathematical continuum field theory of how a relative motion of the myriad protein filaments within axoplasm could drive the cytoplasmic streaming responsible for axoplasmic transport).

Axoplasmic transport is a phenomenon with substantial intrinsic importance, but the larger issue is to understand the mechanism that endows cytoplasm in general with the ability to move. A number of biological energy transduction mechanisms are now understood, at least in broad outline, such as muscle, the bacterial flagella driven by rotary engines, "9+2" eucaryotic flagella, and cilia. The machinery responsible for cytoplasmic motion is not on this list.

I presume too much to refer to <u>the</u> machinery of cytoplasmic motion; very probably there is a large catalog of various such machines. But biological evolution tends to conserve good ideas, and to invent new machinery by tuning up and modifying old machinery. Very possibly then, the entire ensemble of biological mechanisms that transduce chemical energy to mechanical motion will be found to consist of very many canonic variations on <u>very few</u> distinct themes. It is not impossible that the molecular machinery that drives axoplasmic transport and the molecular machinery that drives cytoplasmic streaming in amoebae are two variations on the same theme. So, while this article will henceforth concentrate upon axoplasmic transport, a short aside (section 5) will suggest how the mechanism I propose for axoplasmic transport, modified slightly, could drive amoeboid cytoplasmic streaming.

2. THE PHENOMENON. Each neuron has a single "output wire" called an axon. This is a long (from a millimeter up to meters long in large mammals) small diameter (from 1 to 100 μm in diameter) cylindrical extension of the neuron cell body (soma), bounded by the same plasma membrane that

bounds the soma. That membrane is exitable. The major and
most familiar function of the typical axon is to conduct the
electrical action potential depolarization wave along the
axon membrane from the soma to the far end of the axon
where the arrival of action potential spikes causes the
release, at the axon tip, of a cloud of submicroscopic
vesicles. The vesicles contain a neurotransmitter chemical
(acetylcholine) that diffuses across the junction between

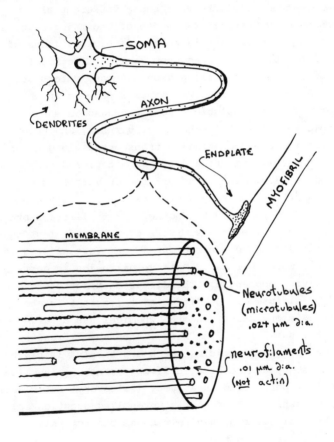

Figure I

the axon tip and the dendrites ("input wires") of another
neuron, or the exitable membrane of a muscle cell, to
initiate depolarization of the membrane of whatever cell the
neuron sends its signals to.

The experimental discovery of how squid axon membranes
conduct the action potential depolarization wave, and the
mathematical characterization of that process using partial
differential equations by Hodgkin and Huxley, and the
subsequent mathematical study of those differential
equations by many researchers is one of the major success
stories of theoretical and mathematical biology (none of
which will be further touched upon here). Electrical
action potentials travel along axons at many meters per
second. 'Axoplasmic transport' names a process in which
many chemicals (some soluble - some packaged in submicroscopic vesicles), mitochondria, viruses, and perhaps
whole cytoplasm are mechanically transported along axons,
both from the cell soma toward the axon tip, and in the
opposite direction. The transport speeds vary from one
animal to another, and vary widely within a single axon,
from one kind of particle to another. But all transport
speeds are about five or six orders of magnitude less than
the electrical propogation speeds; a very fast axoplasmic
speed is one meter per day. Passive diffusion is certainly
not the driving mechanism. The diffusion time for a soluble
chemical of diffusivity D, diffusing a length L is about
L^2/D. If L is one meter, and the chemical has a large
diffusivity, say 5×10^{-6} cm^2/sec., then the diffusion time
is about 83 years! Particles which are barely visible by
light microscopy can be seen in living axons zipping along
at about 1 μm per second in a jumpy (saltatory) but clearly
non-random fashion. A presently unresolved question is:
what is the molecular mechanism responsible for this
dramatic cytoplasmic motion?

Axoplasmic transport plays essential roles in the function of an axon. The cell nucleus, along with the cellular machinery for the manufacture of proteins and other chemicals, reside in the neuron soma. Proteins and other substances made there must be transported along the whole length of the axon. Neurotransmitter chemical, for example, is manufactured there and packaged in vesicles. The black dots strung along the axon in Figure I represent neurotransmitter vesicles, made in the soma, which are transported down the axon to the endplate where they are held for release upon arrival of action potential spikes. Mitochondria, essential suppliers of ATP needed to support the exitability of the axon membrane, etc. are transported. A complete catalog of all important substances transported along axons has not yet been established, but already the list is very long. The next few paragraphs collect selected data from the large experimental literature on axoplasmic transport. See also review articles by Kerkut (1975); Jeffrey and Austin (1973); Kristensson and Olsson (1973); Wuerker and Kirkpatrick (1972).

It is not yet agreed whether various materials are transported in the forward direction (from soma toward axon tip) at various distinct speeds, or whether there is a continuum of possible forward speeds. There is agreement, however, that there are two distinctly different ranges of speeds. In almost all axons, the majority (about 5/6) of all protein transported forward is carried at the so called slow rate of about 1/2 mm. per day. Of this slowly transported protein, about 65% is in soluble form. About 1/6 of the protein transported forward is carried at much faster speeds, which vary from 100 to 1500 mm/day. About 85% of this rapidly transported protein is particulate, that is, packaged in submicroscopic vesicles. Some researchers report a single fast transport speed [see, for example, McEwen, Forman, and Grafstein (1971); McEwen and Grafstein (1968); Fernandez, Burton, and Samson (1971); Cancalon and Beidler (1975)], some report two, or three distinct fast

forward transport speeds [see, for example Karlsson and
Sjostrand (1972); Partlow, et. al. (1972); Norstrom and
Sjostrand (1971); Schonbach and Cuenod (1971); Karlsson
and Sjostrand (1971); Jarrot and Geffen (1972)], while some
report a continuum of transport speeds [see for example,
Bradley, Murchison, and Day (1971)]. Most of these studies
measure transport speeds by injection near the cell somas
of radioactively labeled amino acids which are picked up by
the neurons, manufactured into proteins, then transported
down the axons. The arrival schedule of radioactively
labeled material at some site toward the axon tip is
measured, and the transport speeds inferred. Some studies
involve direct visual evidence [see, for example, Kirkpatrick,
Bray and Palmer (1972); Cooper and Smith (1974); Pomerat,
et. al., (1967)].

Some examples of materials transported from soma to axon
tip are: neurotransmitter vesicles (acetylcholine),
acetylcholinesterase, epinephrine, norepinephrine, RNA,
mitochondria, monoamine oxidase, dopamine β-hydroxylase,
phospholipids, many proteins and glycoproteins, amino acids
(the amino acids mostly in soluble form at the slow speed).

There is a retrograde transport (from axon tip toward
soma) of various materials occuring simultaneously with the
forward transport in the same axons. No slow (1/2 mm./day)
retrograde transport has been reported, and usually the
retrograde transport is reported to proceed at about one-
half of the fastest fast forward transport speed. For
reports specifically dealing with retrograde transport, see
Cooper and Smith (1974); Rannish and Ochs (1972); LaVail
and LaVail (1972); McLean and Burnstock (1972); Kristensson,
Olsson, and Sjostrand (1971); Bray, Kon, and Breckenridge
(1971). Some examples of material known to be transported
retrograde are: tetanus neurotoxin [see Price, et. al.
(1975) for the answer to the long-standing question of how
this toxin reaches the CNS to cause inhibition of synaptic
inputs to motorneurons, thence the clinical symptoms of
"lockjaw"]; nerve growth factor [see Hendry et. al. (1974)];

mitochondria, serum albumin [see Kristensson and Olsson
(1973)]; rabies virus [see Field and Hill (1974)]; horse-
radish peroxidase (for which there is a sensitive assay;
the trajectories of nerve tracts are now being mapped by
seeing where peroxidase, injected locally, gets transported)
[see LaVail and LaVail (1972); Ellison and Clark (1975);
Kristensson and Olsson (1971)].

What molecular machinery drives axoplasmic transport?
Electron micrographs of nerve axons show that a large
number of long protein filaments are distributed throughout
axoplasm all lying with their lengths oriented parallel to
the axon length as depicted in the insert in Figure I. The
filaments seem to be of two types: neurotubules, which are
probably identical with microtubules, which have a diameter
of about .024 µm, and an undetermined length - at least
several µm; and neurofilaments, with a diameter of about
.01 µm and an undetermined length. These protein filaments
occupy perhaps 5% of the volume of the axon [see Peters
(1968); Wuerker and Kirkpatrick (1972); Schmitt (1968)]. In
larger axons, the neurofilaments are concentrated near the
core of the axon, while the neurotubules are concentrated
between the core and the axon membrane, as indicated in
the insert of Figure I. Electron microscope autoradiography
indicates that radioactively labeled proteins rapidly
transported in the forward direction are found mostly in
the zone occupied by the neurotubules [see Lentz (1973);
Schonbach, Schonbach, and Cuenod (1971)].

Very little is known about the chemistry, structure, or
function of neurofilaments, except that they are not the
same protein as actin. Microtubules are ubiquitous in
biological cells -- they constitute the major elements of
flagella, of the mitotic spindle, of the internal skeleton
of many cells -- and a great deal is known about them. In
particular, it is fairly certain that the chemicals
Colchicine, Vinblastine, and Vincristine bind to the subunits
of microtubules and interfere with the assembly and function
of microtubules. Colchicine binds as well to the subunits

of neurotubules, and very probably, neurotubules are in fact,
microtubules that happen to lie in axoplasm. Also Lidocaine,
Methyl Mercury and low temperature disrupt microtubule
function. All of these agents block fast axoplasmic
transport and therefore implicate neurotubules as having a
substantial role in the driving mechanism of axoplasmic
transport [see, for example, Abe, et al (1975); Gross (1973);
Karlsson, Hansson, and Sjostrand (1971); Norstrom, Hansson,
and Sjostrand (1971); James et al (1970), Fink, et al (1972),
Banks, et al (1971)]. Actin- and Myosin-like proteins have
been isolated from axoplasm [Bray (1974)].

Agents that interfere with the production or splitting
of ATP block fast transport, for example: Dinitrophenol,
Cyanide, and anoxia. This suggests that, whatever the
axoplasmic transport engine is, its fuel is, not surpris-
ingly, ATP [see Kerkut   (1975)].

Two studies have shown that Calcium ion, Ca^{++}, may play
a role in regulating axoplasmic transport [see Ochs and
Worth (1975); Hammerschlag, Dravid, and Chiu (1975)]. This
is interesting because Ca^{++} is known to modulate many kinds
of biological motion transducers, for example the acto-
myosin system in muscle and in amoeboid cytoplasm.

Since axoplasmic transport is a process necessary for
the support of the excitable membrane along which electrical
action potential waves course, it is an appealing idea to
hypothesize that, somehow, the conduction of action potentials directly causes axoplasmic transport. A number of
experimental studies show that this is not the case,
however [see, for example Grafstein, Murray, and Ingoglia
(1972); Jankowsha, Lubinska, and Neimierko (1969); Ochs and
Hollingsworth (1971)].

Some experimental· investigators feel that the smooth
endoplasmic reticulum in nerve axons is formed into a
manifold of tubules that run parallel to the axon length
continuously from the soma to the axon tip and act as
conduits or guide channels within which material is transported [see, for example Droz, Rambourg, and Koenig (1975)].

Also, it is possible that the meshwork of neurotubules that permeate axoplasm is arranged so that bundles of neurotubules form guide channels through which vesicles move without being able to escape laterally.

For completeness, I include a citation of the major piece of practical good news that has come from the experimental study of axoplasmic transport. Continuous administration of ethyl alcohol for three weeks, in moderate or even acute dosage, substantially speeds up the fast axoplasmic transport of mitochondria and vesicles (neither slow nor fast transport of soluble material is affected, but you can't have everything). The study [see Israel, Kuriyama, and Yoshikawa (1975)] was done with rats, but it is common practice to extrapolate pharmacological studies from rats to men. The temperance league will insist that this news be printed with the warning that it is not known whether faster transport of vesicles and mitochondria is beneficial or harmful.

To summarize, axoplasmic transport is a process in which some portion of the axoplasm moves from the neuron soma to the axon tip while, simultaneously, some portion of the axoplasm moves in the opposite direction. Various widely varying speeds are seen in the same axon. Simple gross mass balance arguments require that very nearly as much axoplasm flows retrograde as forward -- otherwise, if most of the axoplasm were flowing in one direction, say forward, at anything like the fast forward transport rate, the entire contents of the neuron soma would be drained away in minutes. Light microscope observations leave no doubt that there is a definite and dramatic organized motion of visible particles - that is, there is more to the phenomenon than some kind of chemical "waves" or active diffusion process of soluble chemicals. Just how much of the axoplasm is in motion is an open question. As will emerge in the next section, some investigators feel that only particular vesicles move through stagnant axoplasm, while other investigators think that rivulets of cytoplasm

are set into a streaming motion and carry with them particulate and soluble material. The most visible molecular
building blocks for the transport machinery are the
neurotubules and/or neurofilaments, the smooth endoplasmic
reticulum, and, possibly actin- and myosin-like proteins.

3. A SURVEY OF THEORIES OF THE DRIVING MECHANISM OF
AXOPLASMIC TRANSPORT. It is none too clear that an interesting role for mathematics can be found in the phenomenon
described above. There may be none, in fact. I will try
below to sketch all the conjectures and theories I know of
that have been advanced in explanation of the phenomenon of
axoplasmic transport. A few require mathematical reasoning
and calculations; most do not. The theory of the driving
mechanism that I will propose in a later section will, I
hope, justify inclusion of this article in a volume titled
"Some Mathematical Questions in Biology" as will the
(potential) mathematical content of a few theories surveyed
in this section. I am spending considerable text on describing the biological phenomenon at hand and non-mathematical
conjectures on how it works partially to highlight the
central difficulty of mathematical biology, namely, that in
each instance of the application of mathematics to biology
the overriding question is: "is there really any important
mathematical question here?" The preponderance of answers
to that question are negative.

Paul Weiss, who discovered the phenomenon of slow
axoplasmic transport, has advanced the theory that the
column of axoplasm within an axon is propelled as a semi-
rigid slug by a peristaltic pump that resides in or around
the axon membrane. He proposes and has experimental
evidence that a wavy geometry propogates along the axon
membrane. Figure II depicts the idea. This is claimed to
drive the entire column of axoplasm at the slow transport
rate of about 1/2 mm per day. [See Weiss (1972) for a
recent exposition of this idea, and Biondi, Levy, and Weiss
(1972) for an investigation of the gross rheological

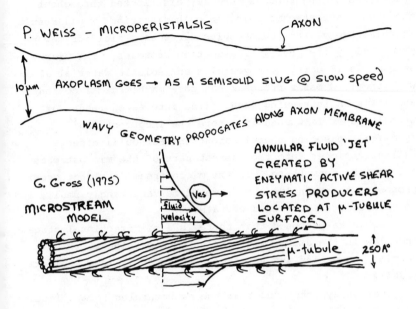

P. WEISS – MICROPERISTALSIS   AXON

10μm   AXOPLASM GOES → AS A SEMISOLID SLUG @ SLOW speed

WAVY GEOMETRY PROPOGATES ALONG AXON MEMBRANE

G. Gross (1975)

MICROSTREAM
MODEL

fluid
velocity   Vas

ANNULAR FLUID 'JET'
CREATED BY
ENZYMATIC ACTIVE SHEAR
STRESS PRODUCERS
LOCATED AT μ-TUBULE
SURFACE

μ-tubule   250A°

Figure II

properties of axoplasm.] One difficulty with Weiss's
microperistalsis theory is the apparent lack of molecular
machinery in the axon membrane to drive the propogation of a
wavy wall geometry. Also, it does not address the question
of what powers fast transport or retrograde transport.

Figure II also depicts an idea proposed by Gross [see
Gross, (1975)], namely that some kind of protein molecules
(such as dynein*) reside along the surface of neurotubule
filaments and can somehow split ATP to create "microstream"
jets of fluid along the lengths of neurotubules.
Neurotubules, if they are essentially the same as microtubules, have an intrinsic polarity. With the directional

*Dynein is the molecule that decorates the microtubules
that constitute "9+2" flagella. A dynein molecule residing
on one pair of microtubule filaments can form attachments
to the neighboring pair of microtubules and, by splitting
ATP, transduce the chemical energy therein to a mechanical
motion of one pair of microtubules sliding past another pair.

polarities of the tubules randomly distributed throughout
the axon, microstreams going in both directions would result.
In this theory, the neurotubules serve as a rigid framework
decorated by the energy transduction proteins. The microstreams would carry both vesicles and soluble material along
with them. Gross's proposal has some mathematical content,
and it is an intriguing idea. (The interested reader of
Gross's paper is warned to ignore the incorrect fluid
dynamical reasoning therein about the Bernoulli effect
pulling vesicles into the fastest part of the microstreams
and keeping them there. In the microstream jets that Gross
proposes, the Reynolds number would be vanishingly small, and
no Bernoulli effect could operate.)

Lubinska proposed that there is a bi-directional streaming flow in the axon, driven by unknown machinery, which
carries organelles and vesicles, etc. with it [see Lubinska
(1964)].

Samson hypothesized that the neurotubules form a lattice
network in the axoplasm column which is covered by some kind
of polyelectrolyte gel that, in a changing ionic environment,
can swell, detach, and inchworm along the supporting
neurotubule lattice, carrying with it vesicles, etc. [see
Samson (1971)].

Hejnowicz suggested that conformational changes propagate
like waves along the intracellular fibrils (neurotubules and
neurofilaments), that these waves have an electrical component, that electrically charged particles could be attracted
toward a fibril and thus come under the longitudinal force
component of the propagating electrical waves and be transported. Hejnowicz presents a substantial body of mathematical
calculation to make the case that, if the electrical waves
were a part of propagating action potentials, his suggested
mechanism could account for the slow transport of protein in
axons [see Hejnowicz (1970)]. He suggests as well that a
turbulent*motion of the fluid caused by the waves might

---

*No fluid motion with miniscule Reynolds number can properly
be called turbulent. Hejnowicz means only that the fluid
motion would consist of disorganized swirls.

provide the basis for activated diffusion. The theory I
shall set forth in the next section involves waves of conformational change propogating along the filaments in
axoplasm, and a swirling motion of the bathing fluid which
transports particulate and soluble material, but does not
involve electrostatic forces acting upon ions.

Lundstrom, in an investigation not addressed to the
problem of axoplasmic transport, showed mathematically that,
under fairly general hypotheses, mechanical waves (both
longitudinal and transverse) on the axon membrane (supposed
to be viscoelastic), with phase velocities corresponding to
action potential propogation velocities, belong to the natural modes of the axon membrane (when the effects of the
surrounding fluid are accounted for). He suggests that
perhaps mechanical vibrations and electrical waves on the
axon membrane are coupled. [see Lundstrom (1974)]. As
memtioned above, the phenomena of action potential conduction and axoplasmic transport seem to be independent. This
citation is made for completeness; if action potential
conduction does excite vibrations of the axolemma, then this
should have some effect upon axoplasmic transport.

From electron microscope studies, Porter has concluded
that some protein material, called, for the time being,
"fuzzy stuff", closely associated with the neurotubules,
forms, with the neurotubules, trabeculae (a manifold of
microchannels) permeating axoplasm. He suggests waves of
structure change in the neurotubule framework, squeezing
material along these trabeculae, as the possible motive
force behind axoplasmic transport. [see Porter (1975)].
In the same vein, Droz, Rumbourg, and Koenig (1975) suggest
that transport occurs inside of a manifold system of smooth
endoplasmic reticulum tubules.

I have collected above, in random order, a number of
proposed theories that involve some kind of gross fluid
motion of all or part of the axoplasm, which carries with it
vesicles and soluble material. Below are assembled those
theories that attribute axoplasmic transport to selfpropelled

and guided vesicles, or selfpropelled "transport filaments".
Fluid motion, if it occurs, is not mentioned, although a
selfpropelled vesicle would certainly create a wake as it
ran through the bathing fluid, and these wakes could enhance
the apparent diffusivity of soluble material.

Schmitt [see Schmitt and Samson (1968)] put forward
a sliding vesicle theory to explain fast transport. His
model, caricatured in Figure III, involves vesicles whose
membranes are decorated by protein molecules (dynein-like?
myosin-like?) which can latch to a neurotubule, split ATP,
and use the energy so released to ratchet the vesicle along,
like a wheel down neurotubule tracks. Since neurotubules
presumably have a directional polarity, both forward and
retrograde transport could be accomplished by these
hypothesized self-propelled vesicles.

I cannot resist making the following remarks in passing:
that enough autonomous ratchets working in concert can
accomplish virtually any task; that no mathematical theoriz-
ing whatever is needed to understand or characterize the
workings of ratchets; that, much to the detriment of
mathematical biology, biological evolution has all too often
opted for ratchets in favor of partial differential equations;
that axoplasmic transport may well be a case in point. Two
more proposed ratchet theories follow.

Ochs suggested a sliding filament model for fast
axoplasmic transport. A protein rod has attached to one side
of it some kind of energy-to-motion transducers (such as
dynein) that can split ATP and slide the rod along microtubules, in much the same way as myosin filaments slide
relative to actin filaments in muscle. Vesicles containing
various materials then attach onto the transport filament
and ride it down the axon. See the caricature of this model
in Figure III. Sometimes, in light microscope observations
of particles (which are usually smaller than a light microscope can accurately resolve) moving in living axons, a row
of several particles is seen to move down the axon as if
they were rigidly bound together in a linear array, and Ochs'
sliding filament model attempts to provide an explanation of
this.

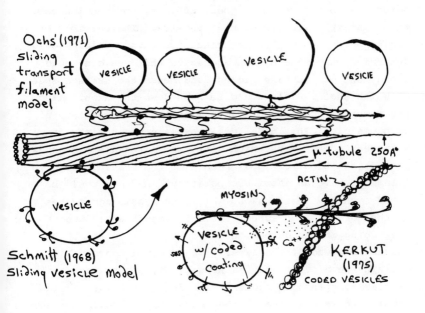

Figure III

The most recently proposed theory is Kerkut's "coded
vesicle" model [see Kerkut (1975)]. Figure III depicts the
way Kerkut proposes that actin filaments are bound to a
microtubule framework. Short myosin molecule aggregates,
dispersed in the cytoplasm, can then slide over these bound
actin filaments. The myosin aggregates occur in a variety
of different lengths, and, presumably, the longer myosin
aggregates, with more myosin heads, would slide down the
axon faster than the shorter ones. In the neuron soma,
different substances are packaged into separate vesicles,
and a variety of antibody-like molecules, bound to the
membranes of these vesicles, code for the contents of the
vesicles. These coding molecules differ, one from another,
in the different binding affinity they have for different
length myosin aggregates. Kerkut's idea is that a chemical
destined for the speediest transport down the axon would be
packaged in a vesicle that has bound to its surface coding

molecules that have the highest affinity for the longest,
thus fastest moving, myosin aggregates. The ensemble of
coding molecules on the membrane of a given vesicle would
determine what size myosin aggregate that vesicle would
spend most of its time riding, and hence the average speed
at which that vesicle would be born down the axon. Also,
the membrane coding could determine the ultimate fate of
each vesicle. This model could explain the variety of
different speeds with which different materials are transported, the bi-directional nature of the transport (actin
filaments have an intrinsic polarity, and some would be
aimed each way in the axon, while the myosin aggregates
in the form of high aspect ratio filaments would tend to
keep going in the direction they started in until they
reached the end of the line...), and could conceivably
explain the above-mentioned sensitivity of the fast
axoplasmic transport machinery to deprivation of Ca^{++}ion.
Namely, it is conceivable that each vesicle could be manufactured in the soma with its own reservoir of Ca^{++}. By
modulating the rate at which it released its supply of Ca^{++},
a vesicle could determine how vigorously the myosin filament
to which it was bound ratchets along the actin threads (in
most acto-myosin systems, the acto-myosin can split ATP
and cause relative sliding only when free Ca^{++} ion is
available to bind to control proteins on the actin filaments)
The study mentioned above [Hammerschlag et al (1975)] of the
calcium sensitivity of axoplasmic transport showed that when
neuron somas are incubated in a Ca^{++} - free medium, fast
transport is inhibited, whereas incubation of the axons
only in Ca^{++} - free medium has no inhibitory effect. This
suggests that some essential Ca^{++} sensitive event must occur
in the soma to initiate and sustain fast transport. This
interpretation is not unambiguous, however, because a host
of experimental studies shows that transport can continue
in segments of axon that have been severed, or otherwise
isolated from the soma.

Again, no task is beyond performance by a multitude of ratchets. If a single set of ratchets won't do, equip each ratchet with an ensemble of ratchets, etc. This is not to detract from the above proposed mechanism of axoplasmic transport, which may be correct. All the theories I have sketched above, and certainly my own, to be described in the next section, suffer from the same malaise. Each is driven by some hypothetical machination of molecules beyond the resolution of current observation techniques. Thus the founding assumptions of each theory are not directly testable.

Before leaving this section, I cite a mathematical investigation which, while not addressed to the phenomenon of axoplasmic transport, may turn out to be relevant. Othmer [see Othmer (1975)] constructed and analyzed a two-phase continuum model of developing tissue possessing a stationary phase (fixed cellular organelles) and a bathing cytoplasm. Exchange of various chemicals can occur by passive diffusion and active transport (the active transport acting in a prefered direction) between the two phases, while the chemicals can undergo reactions locally. Othmer shows that, under fairly general assumptions on the kinds of reactions occuring and the nature of the active transport, a one dimensional stripe of this two-phase continuum can act as a chemical transmission line -- that is, it can propagate finite amplitude chemical waves over long distances at constant velocity with unchanging shape. This kind of phenomenon could be a substantial part of the phenomenon of axoplasmic transport. As mentioned above, there is more to axoplasmic transport than propagation of chemical waves -- there is a coherent observable mechanical motion of particles. But, given a mechanism that moves material mechanically in a preferred direction, which could be interpreted as an active transport mechanism in Othmer's theory, and supposing that the transported material is involved in local chemical reactions, nerve axons could well operate as chemical transmission lines as characterized by Othmer.

4.   A CONTINUUM FIELD THEORY OF THE DRIVING MECHANISM
OF AXOPLASMIC TRANSPORT*.   We suppose that an axon consists
of cytoplasm confined in a rigid tube.   The cytoplasm is
modeled as a mixture of a viscous fluid bathing a meshwork
of rod-like filaments, as depicted in Figure IV.   The
filaments are to represent the neurotubules or neurofilaments
described above, or both acting in concert.   Thus a typical

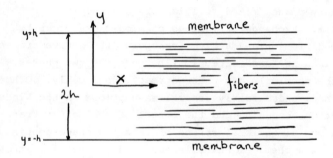

Figure IV

Figure IV.   Schematic representation of an axon
segment.   The space between the rod-like protein
fibers, bounded by the rigid plane boundaries is
filled with viscous water-like fluid.   The plane-
boundary axon caricature of cylindrical axons is
employed to avoid Bessel functions.

---

*This section, and sections 6 and 7, are based largely upon
a paper by Odell (1976b).

filament has a diameter between .01 and .025 μm, and a
length in excess of several μm. The filament constituent
of the mixture will be assumed to occupy a volume fraction
of about 5%. Since the fibers are so small, even at this
low volume fraction, they are densely distributed in the
bathing viscous fluid portion of the cytoplasm, and since,
for any conceivable flow of the fluid relative to the
filaments, the Reynolds number of the flow will be vanishingly small, the meshwork of filaments will be taken to be
a porous medium filled with viscous fluid. It will be an
anisotropic porous medium because all the filaments are
assumed to lie with their lengths oriented parallel to the
axon length.

The point to be made here is simple. If the porous
meshwork of filaments moves, then the fluid will also move.
If the filaments move in an appropriate fashion, then the
fluid can be driven to execute a coherent bidirectional
streaming motion that could carry particulate and soluble
material with it in such a way as to account for many
observed features of axoplasmic transport.

To make this point more precisely, I shall suggest
equations to govern the motion of the fluid constituent of
the cytoplasm given the motion of the filaments, then
postulate a particular kind of motion for the filaments to
execute, then derive the velocity field induced in the fluid.
Next, I will find the trajectories of particles riding in
the fluid to see how the fluid flow would transport non-
diffusible material. Finally, I shall give a highly
simplified discussion of the fate of soluble, hence diffusible, substances carried by the derived fluid flow. Along
the way, I shall make the aside that sketches how the same
mechanism of filaments moving a bathing fluid could characterize the active cytoplasmic streaming by which amoebae
extend pseudopodia.

In this continuum theory, we keep track of the following fields depending upon position $\underline{x}$ and time t (once
underlined quantities are vectors, twice underlined

quantities are second order tensors, quantities not underlined are scalar):

$\underline{u}(\underline{x},t)$ = the fluid velocity field,

$\underline{w}(\underline{x},t)$ = the fiber velocity field,

$f(\underline{x},t)$ = the local fiber volume fraction (with average
value = 5%),

$P(\underline{x},t)$ = the hydrostatic pressure in the fluid,

$\underline{e}(\underline{x},t)$ = the local orientation field of the fibers
(which, in this application, will be always
parallel to the axon length).

Assuming that, in the cytoplasm model, whatever is
not fibers is fluid, that fluid and fibers are incompressible,
and that neither is created or destroyed, we have the two
balance laws:

conservation of fibers: $\dfrac{\partial f}{\partial t} + \text{div}(f\underline{w}) = 0$ (4.1)

conservation of fluid: $\dfrac{\partial}{\partial t}(1-f) + \text{div}[(1-f)\,\underline{u}] = 0$ (4.2)

Each of these is of the form:

$$\left\{ \begin{array}{l} \text{time rate of} \\ \text{change of item} \end{array} \right\} = -\text{div} \left\{ \begin{array}{l} \text{flux of} \\ \text{item} \end{array} \right\} \quad ,$$

where, in (4.1), "item" is proportional to the amount of
fibers per unit volume of space and, in (4.2), item is
proportional to the amount of fluid per unit volume of space.

Instead of a real momentum balance equation, we use
Darcy's law, generalized for a porous medium in which the
substrate moves, namely:

$$\underline{u} = \underline{w} - \frac{1}{\mu}\,\underline{\underline{K}}(f,\underline{e})\ \text{grad } P \tag{4.3}$$

Here, μ is the viscosity of the fluid constituent and $\underline{\underline{K}}(f,\underline{e})$
is the permeability tensor for the anisotropic porous
medium. $\underline{\underline{K}}$ tabulates the permeability of the meshwork of
protein filaments, in different directions. A discussion
of the meaning of equation (4.3) and the tensor $\underline{\underline{K}}$ therein
together with a calculation of the values of the components of
$\underline{\underline{K}}$ can be found in Odell (1976a)[Section 10 and 11].

Essentially, (4.3) says that when the fibers move, the fluid follows the fibers, unless the hydrostatic pressure gradient, modulated by the anisotropic permeability of the fiber meshwork, pushes it elsewhere.

In the reference just cited it is shown that, relative to an orthonormal coordinate system $(\underline{i}_1, \underline{i}_2, \underline{i}_3)$ in which $\underline{i}_1$ is parallel to $\underline{e}$, the fiber orientation direction, and $\underline{i}_2$ and $\underline{i}_3$ are perpendicular to $\underline{i}_1$, the components of $\underline{\underline{K}}(f, \underline{e})$ are (approximately):

$$\underline{\underline{K}}(f, \underline{e})_{ij} = .1f^{-4/3} a^2 \begin{pmatrix} 1 & 0 & 0 \\ 0 & 2/3 & 0 \\ 0 & 0 & 2/3 \end{pmatrix} \qquad (4.4)$$

where a denotes the radius of a typical fiber. This is a data fit to asymptotically calculated values good in the limit as the ratio of fiber diameter to distance between fiber axes goes to zero.

It is easier for fluid to ooze through a universe of oriented fibers in a direction parallel to the fibers (the $\underline{i}_1$ direction) than transverse (the $\underline{i}_2$ and $\underline{i}_3$ directions), and this intuitive property of anisotropic media shows up clearly in (4.4).

If the motion of the fibers is prescribed, the set of equations I have assembled already, together with appropriate boundary conditions, are sufficient to determine the fluid velocity field. To produce a theory for the driving mechanism for axoplasmic transport, I shall specify, as an outright guess, how the fibers move, then deduce how the fluid moves. First, however, I want to indicate how the equations set down so far here, supplemented by balance laws modeling a chemical control of the fiber motion, can characterize how amoeboid pseudopodium extension proceeds.

5.   AN ASIDE: AMOEBOID PSEUDOPODIUM EXTENSION.

An amoeba is a single-celled protozoan animal, which, crudely described, is a tiny bag of amorphous cytoplasm capable of coherent locomotion. It moves by extending a pseudopodium, that is, it extrudes a portion of its cytoplasm to form an exploratory tube-like foot. If this temporary appendage finds something it likes, the whole amoeba can follow the pseudopodium by flowing into it. For a long time, biologists have argued about the mechanism by which amoeba extend pseudopodia. There is general agreement that the outermost cytoplasm of an amoeba exists in a gelled state, while the innermost cytoplasm is in a fluid state capable of streaming. The standard textbook theory of how an amoeba moves has it that the gel layer surrounding the main cell body contracts, raising the hydrostatic pressure there. A part of the periphery weakens and cytoplasm is squeezed out to form the tube-like extension. As the tube forms, the outermost layer of cytoplasm gels.

Robert D. Allen, a cell biologist, convinced H.L. Frisch and myself that the theory just sketched was altogether wrong, and that the extending pseudopodium is, instead, pulled out from the tip. A series of experimental studies by Allen and his coworkers left little doubt that: the machinery responsible for the fountain-like pattern of cytoplasmic streaming existed everywhere dispersed in the cytoplasm, that severed or otherwise isolated parts of that machinery could function autonomously (hence there was no need for a contraction of the cell border and the hydrostatic pressure excess it was supposed to generate), that amoeboid cytoplasm had dispersed in it actin and myosin protein filaments closely analagous to the versions of those molecules that power skeletal and smooth muscle in higher animals, and that, when sufficient ATP was present, the local availability of Ca^{++} ion controlled the local rate at which amoeboid cytoplasm could "contract", presumably by determining how vigorously actin and myosin filaments slide past each other.

When this phenomenon came to my attention, I had been setting up the model of axoplasmic transport to which this article is devoted, and H.L. Frisch and I modified the equations and ideas set forth above to construct a model of amoeboid pseudopodium extension [see Odell and Frisch (1975)].

Intuitively, the model works as follows. A forming pseudopodium is thought of as a rigid outer gel tube full of cytoplasm in a fluid state. This cytoplasm has dispersed in it actin and myosin filaments, all arranged with their lengths parallel to the axis of the pseudopodium. At the tip of the pseudopodium, we suppose that Ca^{++} is either released from some internal storage facility or admitted from the external medium. This Ca^{++} ion diffuses in the fluid portion of the cytoplasm, and it binds to the filaments. When a sufficient amount of Ca^{++} is bound per filament, the filaments, previously disaggregated, start to aggregate together, and to "contract" (actually, the myosin filaments cause the actin filaments to slide past each other. Viewed from a distance, this relative sliding looks like contraction of long strands of aggregated protein filaments.). When the amount of Ca^{++} bound to each filament exceeds a certain threshold, we suppose that the acto-myosin system forms a rigid gel. With the source of the Ca^{++} trigger chemical at the tip of an advancing psuedopodium, this gel naturally forms at the tip. The strands of aggregated filaments are assumed to be attached to this gel at the tip. Ca^{++}, diffusing back from the tip, elicits a contraction of the filaments distributed all along their aggregated lengths, back to the site where aggregation first occured. When the contracting strands pull forward to the tip, they exert a viscous drag upon the bathing fluid and try to drag

it forward (refer to Figure V). The fluid will move forward
with the fibers unless impeded by a sufficiently strong
adverse pressure gradient (with pressure highest toward the
tip). We assumed that the excess of hydrostatic pressure
just inside the pseudopodium tip could roll out the gel
forming at the tip to extend the pseudopodium, and we
assumed there was a linear relationship between the gross
forward speed of the extending psuedopodium and the pressure
drop from the inside of the tip to the ambient external
fluid. The column of cytoplasm (modeled as a porous medium
of contracting fibers bathed by viscous fluid) moves
forward, pulled by the contracting filaments at that speed
which satisfies the above linear relationship.

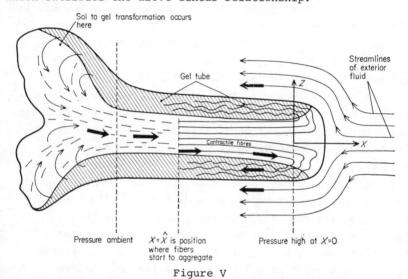

Figure V

Figure V. Schematic representation of amoeboid pseudopodium extension, as seen from a moving coordinate
system attached to the tip of the psuedopodium. The
central column of fluid cytoplasm is seen to be pulled
forward toward the tip by the contracting fibrils
while the forming rigid gel tube appears to move
backwards. The external medium streams past. Some
mechanism must dissolve the gel at the rear end to
recruit new fluid cytoplasm to be pulled up to the tip.*

---

*Reprinted from "A Continuum Theory of the Mechanics of Amoeboid Pseudo-
podium Extension" by Odell and Frisch, Journal of Theoretical Biology 50
(1975), p. 68, with permission of Academic Press.

If Ca^{++} is released at a fixed rate, the extension process self limits at a speed determined by the release rate. As Ca^{++} moves back from the tip by diffusion in the fluid, it is simultaneously convected forward toward the tip by the streaming motion which the Ca^{++} elicits by triggering aggregation and "contraction" of the actin-myosin filaments.

To mathematize the above assumptions, a balance law for Ca^{++} is needed to express the way Ca^{++} diffuses, is convected, and is "destroyed" by a first-order reaction in which Ca^{++} binds to fibers to produce shortened fibers and a fiberbound version of Ca^{++}, which is called B. Another balance law is needed for B, which does not diffuse, but is convected at the local fiber velocity, and is produced in the same measure as Ca^{++} is "destroyed".

A complete description of this model of amoeboid pseudopodium extension, and of the biological phenomenon, can be found in Odell (1976a) or in briefer form, in Odell and Frisch (1975). We return now to the task at hand.

6. DETERMINATION OF THE FLUID VELOCITY FIELD FROM A GUESS
    OF HOW THE FIBERS MOVE.

I will make an outright guess about how the fibers move, then show that the resulting fluid motion could carry material in such a way as to explain many observed features of axoplasmic transport. This constitutes merely a plausibility argument; a single example suffices. The price for entertaining completely general classes of fiber motion would be substantial complication for which no comensurate payoff is in sight.

First, we restrict attention to filament motions brought about by filaments sliding back and forth, parallel to $\underline{e}$, the fiber orientation direction, which we shall take to be the x-axis (see Figure IV), which is parallel to the axon length. Thus $\underline{w}$ has only an x-component, w. Second, we make a two-dimensional axon model in which the axon membrane is taken to be two infinite parallel plane sheets (we avoid circular cylinder geometry in order to avoid Bessel functions). The plane membranes are located at y=h and y=-h. With these

restrictions, I can specify $f(x,y,t)$ or $w(x,y,t)$ and deduce
the other from equation (4.1). It is easiest to specify f
and deduce w. To obtain w, it is necessary to require no
net transport of fibers in either direction along the axon
(or some equivalent assumption). That is, we imagine the
fibers to oscillate in the x-direction around some fixed
base point.

We shall take f to be:

$$f(x,y,t) = f_o [1 + \varepsilon\phi(y,x-ct)]. \qquad (6.1)$$

With $\varepsilon << 1$, and ϕ periodic in x-ct, this represents a
traveling wave perturbation of the local fiber volume
fraction with small amplitude measured by ε, and wave speed
c. We shall leave unspecified how this hypothetical wave of
fiber sliding propogates along the protein filament meshwork.

With f specified, the balance equations (4.1), (4.2),
and (4.3), with $\underline{K}(f,\underline{e})$ given by (4.4), are four linear
partial differential equations of first order in the four
unknown fields w, u_x, u_y, and P, each of which depend upon
x,y, and t. When w is deduced from (4.1), so f and w are
known, the divergence of (4.3), using (4.2) to eliminate
$\underline{u}$, yields a single linear second order p.d.e. for pressure P,
whose variable coefficients are determined by f specified in
(6.1). Boundary conditions must be imposed, and consist of
the constraints that the fluid not penetrate the membranes
($u_y = 0$ at $y = \pm h$), which shows up as $\frac{\partial P}{\partial y} = 0$ at $y = \pm h$, and
that, on average, there is no net fluid transport in either
direction, forward or retrograde. (This last constraint
could take the alternate form of allowing a prescribed net
flow rate of axoplasm down the axon at the universal slow
transport speed of 1/2 mm/day.)

The resulting second-order linear p.d.e., with its
boundary conditions, is solved by a regular perturbation
expansion of P as a power series in the small parameter ε.
In this expansion, f_o, the average fiber volume fraction,
was taken to be of order ε^2. As a particular choice we
take ϕ in (6.1) to be:

$$\phi(y,x-ct) = \frac{1}{2}\left[1 - \cos\frac{\pi y}{h}\right] \sin \hat{K}(x-ct) \qquad (6.2)$$

which models the case where the most vigorous fiber contraction occurs nearest the membrane, or to the case where
we assume that the "contractility" resides mainly in the
neurotubules and the neurotubules are concentrated near the
membranes while the neurofilaments fill the central region
[the case seen experimentally by Lentz(1972)]. $\hat{K}$ is the
wave number of the traveling wave.

The calculation of the fluid velocity field given the
above information is not a difficult task, merely a complicated task. Fortunately it is the kind of task that has
been more or less automated by applied mathematicians and
can be accomplished by "turning a crank" as illustrated in
Figure VI.

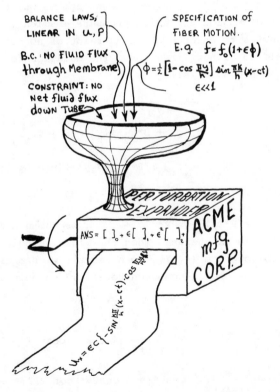

Figure VI.   The regular perturbation expansion solution
method.

To give a few details, we find the first several terms for the series expansion of P in powers of ε. For the actual calculation all variables must be made dimensionless. When we let astirisks denote dimensionless variables and:

$$y = \frac{h}{\pi} y^* \quad \text{and} \quad x = \frac{\beta h}{\pi} x^*, \quad \text{where } \beta^2 = \frac{3}{2},$$

and $\tau^* \equiv \frac{\pi}{\beta h}(x-ct) =$ the dimensionless wave variable, and $f_o = n\varepsilon^2$ where n is of order 1, and

$$P(x,y,t) = P_o[P^*(y^*,\tau^*) + G^*(x^*)] \tag{6.3}$$

where $P_o = \dfrac{\mu h \varepsilon c f_o^{4/3}}{(\cdot 1)\pi a^2} = \dfrac{\mu h c n^{4/3}}{(.1)\pi a^2} \varepsilon^{11/3}$,

we find, first that $\dfrac{dG^*}{dx^*}$ must be a constant, m, which is the time-average dimensionless axial pressure gradient. This must be adjusted to enforce the constraint of no net average flux of fluid in either direction. The equation for the dimensionless pressure is:

$$P_{,\tau^*\tau^*} + P^*_{,y^*y^*} = \varepsilon^0[\phi_{,\tau^*}]$$
$$+ \varepsilon\left[\frac{4}{3}(P^*_{,\tau^*} + m)\phi_{,\tau^*} + P^*_{,y^*}\phi_{,y^*} - \frac{2}{3}\phi\,\phi_{,\tau^*}\right]$$
$$+ \varepsilon^2\left[-\frac{4}{3}\phi\{(P^*_{,\tau^*}+m)\phi_{,\tau^*} + P^*_{,y^*}\phi^*_{,y^*}\}+(n - \frac{5\phi^2}{9})\phi_{,\tau^*}\right].$$
$$\tag{6.4}$$

With $P^* = \varepsilon^0 P^{*(0)} + \varepsilon^1 P^{*(1)} + \ldots$, we compute the first two terms of this expansion. Then the fluid velocity components follow from (4.3). To order ε^2, with ϕ given by (6.2), the dimensioned x-component of fluid velocity, u_x, is given by:

$$\frac{u_x}{c} = -\varepsilon\left[\frac{\sin k^*\tau^* \cos y^*}{2(1 + k^{*2})}\right] + \frac{\varepsilon^2}{48(1+k^{*2})}\left[4(1-k^{*2})\cos y^*+(k^{*2}-3)\cos 2y^*\right]$$
$$+ \frac{\varepsilon^2}{48(1+k^{*2})}\left[-\frac{4(1+3k^{*2})}{(1+4k^{*2})}\cos y^* + 3\cos 2y^*\right]\cos 2k^*\tau^* + O(\varepsilon^3)$$
$$\tag{6.5}$$

The dimensioned y component of fluid velocity is given by:

$$\frac{u_y}{c} = \varepsilon \left[\frac{k^*}{2(1+k^{*2})} \right] \sin y^* \cos k^*\tau^* \ +$$

$$\frac{\varepsilon^2}{48(1+k^{*2})} \left\{ F(\tau^*) + k^* \left[\frac{8(1+3k^{*2})\sin y^*}{1+4k^{*2}} - 3 \sin 2y^* \right] \sin 2k^*\tau^* \right\}$$

$$+ O(\varepsilon^3) \qquad\qquad (6.6)$$

where $F(\tau^*)$ is periodic, and inconsequential in later calculations. In (6.5) and (6.6), k^* is the dimensionless wave number:

$$k^* = \beta h \hat{k}/\pi$$

7. TRAJECTORIES OF NON-DIFFUSIBLE PARTICLES.

In equations (6.5) and (6.6), the fluid velocity field is seen to be periodic in space and time. A closer examination will now reveal that, at a particular height, y^*, while the fluid goes back and forth, it goes a little more forth than back. At other heights, there is more back than forth to the fluid motion, and in this way, the fiber motion hypothesized above drives a flow which, on average, is a bi-directional flow. There is a subtlety to this point. If an observer were to fix his attention upon a particular position in space and watch the fluid swirling according to (6.5) and (6.6), he would (rightly) conclude that the time average of the x-component of velocity at height y^* is equal to the second term of equation (6.5), namely

$$[\text{time-average value of } u_x \text{ at } y^*] =$$

$$\frac{\varepsilon^2 c}{48(1+k^{*2})} \left[4(1-k^{*2})\cos y^* + (k^{*2}-3)\cos 2y^* \right] .$$

It does not follow that this is the average drift speed accomplished by a particle riding in the fluid flow with average height y^*.

Let $(X(t), Y(t))$ be the coordinates, at time t, of a fluid particle which, at time $t=0$, was located at $(0, y_0)$,

that is, $(X(t),Y(t))$ are the LaGrangian coordinates of a
material particle. To find the trajectory of this particle,
we must solve this set of non-linear differential equations:

$$\frac{dX}{dt} = u_x(X(t),Y(t),t)$$

$$\frac{dY}{dt} = u_y(X(t),Y(t),t) \qquad (7.1)$$

when the velocity field $u_x(x,y,t)$, $u_y(x,y,t)$, is prescribed,
subject to the initial conditions: $X(0) = 0$ and $Y(0) = y_o$.

Equations (7.1) say that the velocity of the particle
whose trajectory we want is the same as the velocity of the
flow field wherever the particle happens to be.

With u_x and u_y given by (6.5) and (6.6), the equations
(7.1) are complicated. To solve them, we use the same machine depicted in Figure VI. Before writing down the
results, let me illustrate how a purely oscillatory flow
field can effect a net drift of fluid particles, a phenomenon familiar to oceanographers, for example, because the
back and forth wave motions in the oceans do transport
particles riding in the water over long distances.

Consider a one-dimensional flow, with only an x-
component to its velocity field, which is purely sinusoidal:

$$u(x,t) = \varepsilon c \, \sin[k(x-ct)] \qquad (7.2)$$

By taking $\varepsilon \ll 1$, we assume that the fluid speed is much less
than the wave speed, c. Solution of

$$dX/dt = u(X(t),t) = \varepsilon c \, \sin[k(X(t)-ct)] \qquad (7.3)$$

will determine the trajectory $X(t)$ of a particle riding in
this flow which was at position 0 at time t=0, if we impose
$X(0)=0$. We can try to solve the nonlinear equation (7.3) by
supposing that, for sufficiently small times, t,

$$X(t) = X_o(t) + \varepsilon \, X_1(t) + \varepsilon^2 \, X_2(t) + \ldots \qquad (7.4)$$

If we substitute (7.4) into (7.3), and use trigono-
metric expansions, we obtain:

$$\dot{X}_o + \varepsilon\dot{X}_1 + \varepsilon^2\dot{X}_2 + O(\varepsilon^3) = \varepsilon c\left\{-\sin kct + \varepsilon k X_1 \cos kct + O(\varepsilon^2)\right\}$$

(7.5)

When we equate each power of ε in (7.5) separately to zero,
and use the initial condition, we find:

$$X_o = 0$$

$$X_1 = \frac{\cos[kct]-1}{k}$$ (7.6)

$$X_2 = \frac{ct}{2} + \{t\text{-periodic terms}\}$$

and hence we have an approximate solution:

$$X(t) = \varepsilon\left\{\frac{\cos[kct]-1}{k}\right\} + \varepsilon^2\left\{\frac{ct}{2}\right\}$$

$$+ \varepsilon^2\{t\text{-periodic terms}\} + O(\varepsilon^3) .$$ (7.7)

Of course (7.7) is valid only for sufficiently small
times since the $\frac{1}{2}\varepsilon^2 ct$ term blows up as $t \to \infty$. But, we see
that, approximately, the particle trajectory consists of an
order ε oscillatory motion on which is superimposed a
steady drift at speed $\frac{1}{2}\varepsilon^2 c$. This simple example illustrates
qualitatively what happens in the axoplasmic flow problem
at hand.

We define the time-average drift velocity profile of
fluid particles whose average height (in Figure IV) is
y_o to be

$$\bar{Q}(y_o) = \lim_{T\to\infty} X(T)/T$$ (7.8)

where $X(t)$ solves (7.1). We know, by symmetry, that there
can be no net drift in the y-direction. To order ε^2, this
drift velocity profile turns out to be, for the case at hand:

$$\bar{Q}(y_o) = \frac{\varepsilon^2 c}{48(1+k^{*2})} \left\{ 4(1-k^{*2})\cos\left(\frac{\pi y_o}{h}\right) + \frac{\left[3-k^{*2}(2-k^{*2})\right]\cos\left(\frac{2\pi y_o}{h}\right)}{1+k^{*2}} \right\}$$

$$(7.9)$$

Note that, as forecast in the beginning of this section, this average drift speed for a particle at average height y_o is _not_ just the steady part of the   flow velocity field exhibited in equation (6.5); there is a substantial difference.

Equation (7.9) gives the drift velocity profile for particles which do not diffuse, but follow the moving fluid exactly. Another case of non-diffusible particles is of interest. As mentioned above, there may be guide channels, composed of bundles of neurotubules or other structural elements in axons, along which particles may move parallel to the axon length without being able to excape laterally. For this reason, it is of interest to compute the time-average drift velocity profile of particles which follow the x-component of the fluid motion, but, due to lateral constraints of one kind or another, cannot move in the y-direction. Thus, we calculate the trajectory X(t), of a particle whose y-coordinate is always y_o, and which starts at $X=0[X(0)=0]$, by solving:

$$\frac{dX}{dt} = u_x(X(t),y_o,t) \qquad (7.10)$$

where u_x is given by (6.5). The result, to order ε^2, is:

$$\bar{U}(y_o) = \frac{\varepsilon^2 c}{48(1+k^{*2})} \left\{ 4(1-k^{*2})\cos\left(\frac{\pi y_o}{h}\right) + \frac{3-k^{*2}(2-k^{*2})\cos\left(\frac{2\pi y_o}{n}\right)}{1+k^{*2}} \right\}$$

$$(7.11)$$

This expression is significantly different from (7.9) by virtue of a removal of the square brackets from the last term in (7.9).

In the above equations there appears a dimensionless

wave number, k*. So far, this parameter has been assigned no definite value. If k* is allowed to vary, a variety of interesting drift velocity profiles, given in (7.9) and (7.11) can be obtained. For one example, when k*=.8725, the drift velocity profile of fluid particles that follow the fluid motion exactly is shown in Figure VII.

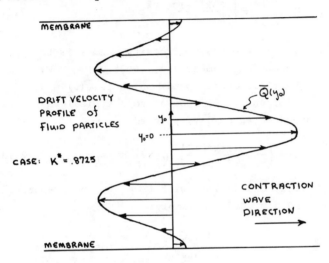

Figure VII
$\bar{Q}(y_o)$, when k*=.8725.

This case, shown in Figure VII, is interesting because, as mentioned in section 2, many experimental investigators of axoplasmic transport report two distinct fast forward transport rates, and a cloud of non-diffusible particles injected into a flow of the sort just analyzed would be born at two quite different speeds; the material at the axis would go to the right much faster than the material near the membranes.

The wave number parameter, k*, cannot be varied with impunity, however. In fact, only one value is likely to be permissable as shown by the following consideration. The time-average axial pressure gradient, which showed up in

equation (6.4) as a parameter, m, must be adjusted to enforce the constraint of no net axial transport of cytoplasm.
When this is done (it is done automatically by the machine
depicted in Figure VI), the time average pressure gradient,
$\frac{\partial P}{\partial x}$, comes out to be:

$$\frac{\overline{\partial P}}{\partial x} \approx \left\{ \frac{\mu f_o^{4/3}}{4.8a^2} \right\} \left[\frac{3k'^2-1}{1+k'^2} \right] \varepsilon^2 c \qquad (7.12)$$

where, again, μ is the fluid viscosity, and a is the radius
of a typical protein filament (say .0075 μm). Now, to make
the maximum particle drift speed $\left(\text{which is } \overline{Q}(0) = \overline{U}(0) = \varepsilon^2 c[7-3k*^4-2k*^2]/48(1+k*^2)^2\right)$ equal to a typical fast forward
transport speed of 40 cm/day, we need

$$\varepsilon^2 c = \left(\frac{(1+k*^2)^2}{7-3k*^4-2k*^2} \right) (.022) \text{ cm/sec} \qquad (7.13)$$

Together, (7.12) and (7.15), with f_o = 5%, and μ appropriate
for water, and a = .0075 μm, imply:

$$\frac{\overline{\partial P}}{\partial x} \approx 1.5 \ \frac{(1+k*^2)(3k*^2-1)}{7-3k*^4-2k*^2} \text{ atmospheres/cm !!}$$
$$(7.14)$$

This makes two important points. First, it shows that
submicroscopic fibers moving in a bathing viscous fluid can
make an extraordinally strong pump. Second, to avoid
catastrophic pressure gradients on the order of an atmos-
phere over several centimeters, the dimensionless wave
number must have the value

$$k* = \sqrt{1/3} \qquad (7.15)$$

This value alone will make the axial pressure gradient
vanish approximately.

Using the above single admissable value of k*, and
ε=.1 in equation (7.13), we find an estimate of the speed, c,
at which the hypothesized waves of fiber sliding must propogate down an axon to cause, by the mechanism analyzed
here, a fast forward transport rate of 40 cm/day, namely

$$c = .65 \text{ cm./sec} = 56160 \text{ cm/day} \qquad (7.16)$$

Figure VIII shows the fluid particle drift velocity profile, $\bar{Q}(y_o)$, as given by equation (7.9) in the critical case $k*^2 = 1/3$. There is a time average drift from left to right (the waves propogate from left to right) in the central core, while there is a (retrograde) drift in the opposite direction near the membranes. The maximum retrograde speed is about one-half the maximum forward speed, and this corresponds to the speeds seen experimentally by Rannish and Ochs (1972).

A study by Cooper and Smith (1974) shows that for each visible particle moving forward in fast axoplasmic transport, ten move in the retrograde direction. This situation could be sustained only if ten particles are made near the axon tip for every one made in the soma. The theory given here has nothing to say about how, where, and in what ratio, particles are made to be transported but addresses the question of how particles, injected into the streaming flow, would be transported.

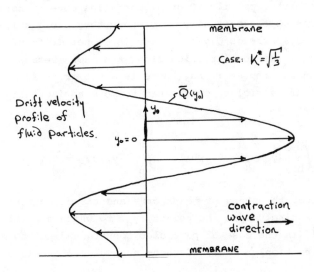

Figure VIII.

176   GARRETT M. ODELL

Figure IX shows the time average drift velocity profile of particles that may be constrained to move only in
the x-direction, given as $\bar{U}(y_o)$ in equation (7.11) for the
same case of $k*^2 = 1/3$.

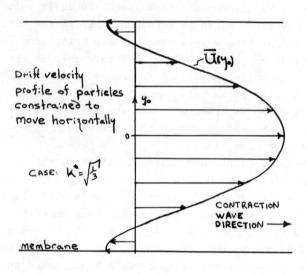

Figure IX.

For either profile, $\bar{Q}$ or $\bar{U}$, particles move in both
directions, but except at $y_o = 0$ where they are equal,
$\bar{U}(y_o)$ always exceeds $\bar{Q}(y_o)$. There is no net drift of
cytoplasm in either direction along the axon (because that
condition was imposed), but there is a net drift from left
to right of particles constrained to move horizontally
only, that is:

$$\int_{-h}^{h} \bar{U}(y_o)dy_o > 0, \text{ while } \int_{-h}^{h} \bar{Q}(y_o)dy_o = 0 .$$

Notice that the speed of transport (and the direction)
depends entirely upon the height, y_o, to which a particle is
confined (in the case of particles constrained not to move in
the y-direction). Thus, if there are guide channels along

axons, the speed and direction at which a particle is transported by the mechanism suggested in this article could be
determined simply by selecting the proper guide channel into
which to insert the particle.

Some experimental correlates of the features of the flow
analyzed herein may be found in Odell (1976b). One significant feature is that the oscillatory flow field analyzed
herein would transport particles in a jumpy back-and-forth
fashion. Light microscope observations of particles undergoing fast axoplasmic transport reveal jumpy or saltatory
motions of the particles.

It must be emphasized that these calculations form not
a genuine theory of the driving mechanism of axoplasmic
transport, but a plausibility argument. Only one illustrative guess of how fibers might move was tried. A host of
other possible guesses suggest themselves, and I would like
to sketch one explicitly. Almost certainly there exists no
simple bi-directional stream of cytoplasm in axons with a
velocity profile as pictured in Figures VIII and IX
(Private communication from Dr. D. Forman). More likely,
there is an ensemble of small rivulets, some going forward,
some going retrograde. By allowing the flow domain in the
problem solved above to extend some number of multiples of
h above and below y=0, a drift velocity profile with a
number of forward and retrograde streams will obviously
result, driven by the same kind of waves of filament sliding
as suggested above with fibers moving only in the direction
of their own lengths.

A different kind of fiber motion, I believe, will result
in an ensemble of forward and retrograde rivulets corresponding to the ideas suggested by Porter cited above (see
Porter 1976). Namely, a case in which fibers move perpendicular to their lengths (that is, perpendicular to the axon
length). Figure X depicts the idea. The solid lines show
typical filaments, with the arrows showing their instantaneous directions of motion. The dashed lines show the
conjectured instantaneous fluid streamlines that would result·

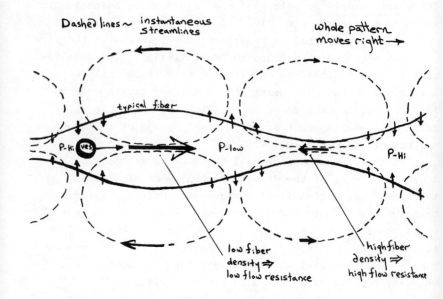

Figure X.

Such a pattern of fiber motion, mirrored into many
layers in the y direction, should result in layers of back
and forth rivulets of fluid, and in addition, yield an
obvious mechanism by which vesicles could be squeezed along
neurotubule guide channels as indicated in Figure X. I have
not yet done the requisite calculations because it is a very
tedious matter to turn the crank of the machine shown in
Figure VI.

The point I hope to have made in this article is that
various kinds of rhythmic waves of fiber motion propogating
along the meshwork of protein filaments in nerve axons can
drive unsteady bi-directional streaming flows of the fluid
bathing the filaments - the kind of flow that could explain
many features observed of axoplasmic transport with, es-

sentially, a single unifying hypothesis.

I have addressed, so far, only the matter of fast forward and retrograde transport. I conclude, in the following section, with some ideas of how the kind of bi-directional flow analyzed above might transport diffusible, soluble
material.

8. THE TRANSPORT OF DIFFUSIBLE MATERIAL.

There is experimental evidence that the material born
by fast axoplasmic transport is largly particulate (hence
of low diffusivity), while most of the material transported
at the slow rate, $\frac{1}{2}$ mm/day, is soluble, hence of much higher
diffusivity [see Sabri and Ochs (1973); Wooten (1973);
Karlsson and Sjostrand (1971b); Sjostrand and Karlsson (1969);
McEwen and Grafstein (1968)]. Diffusible material would be
carried by the sort of bi-directional oscillatory flow
suggested herein more slowly than non-diffusible material.
The reason is intuitively clear. Diffusing particles never
stay for very long at the same y-level, but keep diffusing
from a region in which fluid drifts to the right into
regions in which fluid drifts to the left. Diffusion
superimposed upon this bi-directional convection will be
more rapid than diffusion in a stagnant medium, but much much
slower than the transport of particles that remain always in
a region where fluid is drifting, say, left to right.

A diffusible material injected as a sharp band at some
site, say the soma, would not be carried,by the kind of
fluid motion analyzed above, as a sharp band, but would spread
as more and more of the material was left behind the diffusion front. Experimental evidence that this does in fact
occur in real axons can be found in two recent studies [see
Gross and Beidler (1975); Cancalon and Beidler (1975)].

The task of solving the convection diffusion equation

$$\frac{\partial c}{\partial t} = D\left(\frac{\partial^2 c}{\partial x^2} + \frac{\partial^2 c}{\partial y^2}\right) - \frac{\partial}{\partial x}(cu_x) - \frac{\partial}{\partial y}(cu_y) \qquad (8.1)$$

where the velocity components, given by (6.5) and (6.6),
which are oscillatory in time and position, is beyond the

patience of this author, even if he had a motor-driven model
of the machine in Figure VI. A simplified calculation tells
a great deal, however. Let us suppose that we have two
steady fluid streams, each of height h, moving at a fixed
speed, u, in opposite directions, as sketched in Figure XI.
Suppose a chemical is carried by each stream, and the
diffusivity of the chemical is D. Suppose that the diffusion
of the chemical, whose concentration in the top, left-moving
stream is $C_1(x)$ in steady state, and whose concentration is
$C_2(x)$ in the lower right-moving stream, is somewhat impeded
by some partial barrier to diffusion separating the two
streams. Thus, the diffusive flux of the chemical in the
streamwise direction is $-D\frac{dc}{dx}$, while the diffusive leakage
from the top stream to the bottom (each stream is assumed to
be well mixed for simplicity) is taken to be:

$$\frac{\gamma D}{h^2} \, [C_1(x) - C_2(x)]$$

at position x, where the smaller γ is, the greater is the
attenuation of cross-stream diffusion. If we suppose that
both $C_1(0)$ and $C_2(0)$ have the same fixed value C_{hi}, while
at the other end of the streams, of length L, $C_1(L)=C_2(L)=C_{low}$,
then we find the concentration fields by solving these two
linear, coupled, second-order, ordinary differential
equations:

$$-u\,\frac{dC_1}{dx} = D\,\frac{d^2C_1}{dx^2} + \frac{\gamma D}{h^2}\,(C_2-C_1)$$

$$u\,\frac{dC_2}{dx} = D\frac{d^2C_2}{dx^2} - \frac{\gamma D}{h^2}\,(C_2-C_1)$$

(8.2)

subject to the boundary conditions just cited.

   When we solve this simple problem, and assume that
h << L, we find this expression for the net flux of the
chemical from x=0 to x=L:

$$\text{Net Flux} = \frac{2h}{L}\,(C_{hi}-C_{low})\left\{D + \frac{h^2u^2}{2\gamma D}\right\}$$

(8.3)

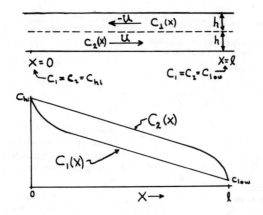

Figure XI.

The concentration fields look as sketched in Figure XI. The effective diffusivity of the chemical is the expression in the curly brackets in (8.3), and this is the item of interest. Experimentally, the speed of slow transport of soluble material seems to be more or less independent of the molecular weight, hence of the diffusivity, of the material transported. The diffusivity of a chemical varies inversely as the cube root of its molecular weight. The expression just derived for the "effective diffusivity" of materials born by a steady bi-directional stream, namely

$$\mathcal{D} \equiv \left\{ D + \frac{h^2 u^2}{2\gamma D} \right\} \qquad (8.4)$$

may suggest how materials of widely differing molecular weights could be born at the same slow transport speed.

182   GARRETT M. ODELL

Figure XII is a plot of effective diffusivity (on a linear
scale) versus molecular diffusivity, D(on a logarithmic
scale), where, for illustration, I have used these values
in (8.4): h = 5 μm; u = 1000 μm/day; γ = .1.

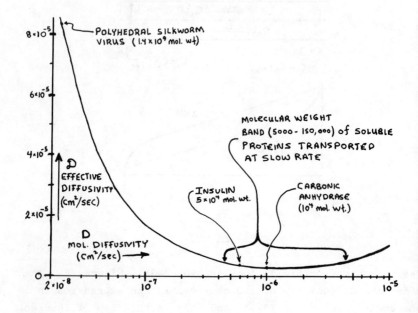

Figure XII.

As can be seen, while the actual diffusivity varies over an
order of magnitude for soluble materials born at the slow
transport rate, the effective diffusivity (using the above
guess parameters) is very nearly constant.

This calculation is merely suggestive of what a
solution to equation (8.1) might show, but it gives some
hint that, with the fiber motion selected so as to produce
a fluid flow field that would characterize fast transport,
both forward and retrograde, by the time-average drift of
non-diffusible particles confined laterally to guide
channels, that same fluid flow field might explain, as well,

slow transport of diffusible material by an effective diffusivity of all soluble material which is independent of
molecular weight due to diffusion and convection in a bi-
directional stream.

ACKNOWLEDGEMENT.

Conversations with Dr. David Forman contributed substantially to my understanding of experimental aspects of
axoplasmic transport as have conversations with Dr. R. D.
Allen. Fine typescript from marginally legible manuscript
is the work of Virginia Steffen.

GARRETT M. ODELL
Department of Mathematical Sciences
RENSSELAER POLYTECHNIC INSTITUTE
Troy, New York, 12181

REFERENCES.

Abe, T., T. Haga, and M. Kurokawa, (1975), Brain Res. 86, (3), p 504-508.

Banks, P., D. Mayor, M. Mitchell, D. Tomlinson, (1971), J. Physiol. (London), 216, (3), p 625-639.

Biondi, R.J., M.J. Levy and P.A. Weiss, (1972), Proc. Natl. Acad. Sci. USA, 69 (7), p. 1732-1736.

Bradley, W.G., D. Murchison, and M.J. Day, (1971), Brain res., 35, p 185-197.

Bray, D., (1974), Endeavor, 33, p 131-136.

Bray, J.J., C.M. Kon and B.M.L. Breckenridge, (1971), Brain Research, 26 (2), p. 385-394.

Cancalon, P. and L.M. Beidler (1975), Brain Res., 89, p. 225-244.

Cooper, P.D., and R.S. Smith (1974), J. Physiol., 242, p. 77-97.

Droz, B., A. Rambourg, and H.L. Koenig, (1975), Brain Res., 93 (1), p 1-14.

Ellison, J.P. and G.M. Clark, (1975), J. Comp. Neurol., 161, (1), p 103-114.

Fernandez, H.L., P.R. Burton, and F.E. Samson, (1971), J. Cell Biol. 51 (1), p. 176-192.

Field, H.J., and T.J. Hill, (1975), J. Gen. Virol., 20, (1), p. 145-148.

Fink, R., Kennedy, Hendrickson, Middough, (1972), Anesthesiology, 36, (5), p. 422-432.

Grafstein, B., M. Murray, and N.A. Ingoglia, (1972), Brain Res., 44 (1), p. 37-48.

Gross, G.W., (1973), Brain Res. 56, p. 359-363.

Gross, G.W., (1975), in International Symposium on the Physiology and Pathology of Dendrites, vol. 12 of Advances in Neurology, G.W. Kreutzberg Ed., Raven Press.

Gross, G.W., and L.M. Beidler (1975), J. of Neurobiology 6, p. 213-232.

Hammerschlag, R., A.R. Dravid, and A.Y. Chiu, (1975), Science, 188, p. 273.

Hejnowicz, Z., (1970), Protoplasma, 71, p. 343-364.

Hendry, T.A., R. Stach, and K. Herrup, (1974), Brain Res., 82, (1), p. 117-128.

Israel, M.A., K. Kuriyama, and K. Yoshikawa, (1975), Neuropharmacology, 14 (5/6), p. 445-452.

James, K.A.C., J.J. Bray, I.G. Morgan and L. Austin, (1970), Biochem. J. 117 (4), p. 767-771.

Jankowska, E., L. Lubinska, and S. Neimierko, (1969), Comp. Biochem. Physiol. 28 (2), p. 907-913.

Jarrott, B. and L.B. Geffen, (1972), Proc. Natl. Acad. Sci. USA, 69 (11), p. 3440-42.

Jeffrey, P.L., and L. Austin, (1973), Prog. Neurobiol., 2, p. 207-255.

Karlsson, J.O., H.A. Hansson, and J. Sjostrand, (1971), Z. Zellforsch Mikroskop ANAT. 115 (2), p. 265-283.

Karlsson, J.O., and J. Sjostrand, (1971a), J. of Neurochem. 18, p. 749-767.

Karlsson, J.O., and J. Sjostrand, (1971b), J. of Neurobiol., 2 (2), p. 135-143.

Karlsson, J.O., and J. Sjostrand, (1972), Brain Res. 47 (1), p. 185-194.

Kerkut, G.A., (1975), Comp. Biochem. Physiol., 51A, p. 701-704.

Kirkpatrick, J.B., J.J. Bray, and S.M. Palmer, (1972), Brain Res., 43 (1), p. 185-194.

Kristensson K., and Y. Olsson (1973), Prog. Neurobiol, 1, p. 95-110.

Kristensson, K., and Y. Olsson, (1971), Acta Neuropatholog, 19 (1), p. 1-9.

Kristensson, K., Y. Olsson, and J. Sjostrand, (1971), Brain Res. 32 (2), p. 399-406.

LaVail., J.H., and M.M. LaVail, (1972), Science (Wash.) 176 (4042) p. 1416-1417.

Lentz, T.L., (1972),J. Cell Biol. 52 (3), p. 719-732.

Lubinska, L., (1956), Exp. Cell Res. 10, p. 40-47.

Lubinska, L. (1964), Prog. Brain Res., 13, p. 1-71.

Lundstrom, I. (1974), J. Theor. Biol., 45, p. 487-499.

McEwen, B.S., and B. Grafstein, (1968), J. Cell Biol., 38 (3) p. 494-508.

McEwen, B.S., D.S. Forman, and B. Grafstein, (1971), J. of Neurobiol. 2 (4), p. 361-377.

McLean, J.R. and G. Burnstock, (1972), Z. Zellforsch Mikroskop Anat. 124 (1), p. 44-56.

Norstrom, A. and J. Sjostrand, (1971), J. Neurochem. 18 (11), p. 2017-2026.

Norstrom, A., H.A. Hansson, and J. Sjostrand, (1971), Z. Zellforsch Mikroskop Anat. 113 (2), p. 271-293.

Ochs, S., and R. Worth (1975), Science (Wash. D.C.) 187, p. 1087-1089.

Ochs, S. and D. Hollingsworth, (1971), J. Neurochem. 18 (1), p. 107-114.

Odell, G.M. (1976a), in Lectures in Applied Mathematics, Vol. 16, Modern Modeling of Continuum Phenomena, R.C. DiPrima, editor, American Mathematical Society, Providence, R.I., in press.

186 GARRETT M. ODELL

Odell, G.M. (1976b), "A Continuum Theory of Axoplasmic Transport", J. Theor. Biology, in Press.

Odell, G.M. and H.L. Frisch, (1975), J. Theor. Biol. 50, p. 59-86, Appendix A.

Othmer, H.G., (1975), J. Mathematical Biol., 2, p. 133-164.

Partlow, L.M., C.D. Ross, R. Motwani, and D.B. McDougal, (1972), J. General Physiology 60(4), p. 388-405.

Peters, A., (1968),The Structure and Function of Nervous Tissue, ed. by G.H. Bourne, Academic Press.

Pomerat, H. (1967), in The Neuron, H. Hyden , editor, pp. 119-178, Elsevier Publ., Amsterdam.

Porter, K. (1967), in Proceedings of the Symposium on Cell Motility (held at the Cold Spring Harbor Laboratory, Sept. 1975), CSH Lab publication.

Price, D.L., J. Griffin, A. Young, K. Peck, A. Stocks, (1975),Science, 188, p. 945 et. seq.

Ranish, N. and S. Ochs, (1972),J. Neurochem. 19 (1), p. 2641-2649.

Sabri, M.I. and S. Ochs, (1973), J. of Neurobiol. 4 (2), p. 145-165.

Samson, F.E., Jr., (1971), J. of Neurobiol. 2 (4), p. 347-360.

Schmitt, F.O., (1968),Proc. Natl. Acad. Sci. 60, p. 1092-1101.

Schmitt, F.O. and F.E. Samson, (1968), (A report based on two NRP conferences held August 16, 1967 and Jan. 15, 1968), Bulletin Neurosciences Res. Prog. 6 p. 113-219 (see page 139 et. seq.).

Schonbach, J., C. Schonbach, and M. Cuenod (1971), J. Comp. Neurol., 141 (4), p. 485-498.

Schonbach, J. and M. Cuenod, (1971), Exp. Brain Res., 12 (3), p. 275-282.

Sjostrand, J. and J.O. Karlsson, (1969),J. Neurochem. 16 (6), p. 833-844.

Weiss, P.A., (1972),Proc. Natl. Acad. Sci. USA, 69 (3), p. 620-623.

Wooten, G.F., (1973), Brain Res. 55, p. 491-494.

Wuerker, R.B., and J.B. Kirkpatrick, (1972), Int. Rev. Cytology, 33, p. 45-70, edited by G.H. Bourne, J.F. Danielli, and K.W. Jean, Academic Press.